Detection of Epizootic Ulcerative Syndrome in Freshwater Edible Fish Using AI Techniques

Authored By

Podeti Koteshwar Rao
Department of Zoology
SVS Degree and PG College
Affilitated to Kakatiya University
Warangal, Telangna, India

Detection of Epizootic Ulcerative Syndrome in Freshwater Edible Fish Using AI Techniques

Author: Podeti Koteshwar Rao

ISBN (Online): 979-8-89881-534-9

ISBN (Print): 979-8-89881-535-6

ISBN (Paperback): 979-8-89881-536-3

Published by Bentham Science Publishers Pte. Ltd. Singapore, in collaboration with Eureka Conferences, USA.

First published in 2026.

Bentham Science Publishers Pte. Ltd.
No. 9 Raffles Place
Office No. 26-01
Singapore 048619
Singapore
Email: subscriptions@benthamscience.net

CONTENTS

PREFACE

Aquaculture is emerging as one of the most vital sectors in global food production, contributing significantly to nutrition, employment, and economic development. Among freshwater resources, edible fish species serve as a crucial protein source for millions of people, particularly in developing countries. However, the sustainability of aquaculture is severely threatened by infectious diseases, of which Epizootic Ulcerative Syndrome (EUS) is among the most destructive. EUS is a trans boundary aquatic disease, characterized by deep ulcerative lesions, extensive tissue necrosis, and high mortality rates in susceptible fish species. Its outbreaks not only cause severe economic losses to farmers but also affect food security and ecosystem health.

Traditionally, the diagnosis of EUS has relied on clinical observation, histopathological examination, and microbiological or molecular techniques. While effective, these methods are often limited by delays, the need for specialized expertise, and high operational costs. In an industry where timely detection is the difference between containment and catastrophic loss, there is an urgent need for innovative, rapid, and reliable diagnostic solutions.

This research introduces Artificial Intelligence (AI) as a transformative tool for the detection of EUS in freshwater edible fish. AI techniques, particularly machine learning, deep learning, and image recognition, offer the ability to analyze complex datasets, identify subtle pathological features, and provide real-time diagnostic support. By applying AI to fish health monitoring, this work seeks to bridge the gap between traditional diagnostic practices and the need for scalable, automated systems that can be deployed even in resource-limited aquaculture settings. The preface to this study emphasizes the interdisciplinary nature of the work. It brings together insights from aquaculture, veterinary pathology, computational biology, and artificial intelligence. Such convergence is vital for developing next-generation solutions that ensure both the economic viability of aquaculture and the safety of food supplies. Moreover, the approach discussed herein demonstrates how AI can be harnessed not only for accurate disease detection but also for predictive surveillance, enabling farmers and policymakers to anticipate and mitigate outbreaks before they escalate.

This endeavor is motivated by the vision of fostering sustainable aquaculture through technological innovation. By combining biological understanding with computational intelligence, the study aspires to contribute toward improved fish health management, reduced economic vulnerability, and strengthened food security. The application of AI in detecting EUS is more than a technological advancement; it represents a paradigm shift towards precision aquaculture, where science and technology work hand in hand to secure the future of freshwater resources.

Podeti Koteshwar Rao
Department of Zoology
SVS Degree and PG College
Affilitated to Kakatiya University
Warangal, Telangna, India

ABSTRACT

Epizootic Ulcerative Syndrome (EUS) is a highly destructive fish disease that affects a wide range of freshwater edible species, causing ulcerative lesions, high mortality, and substantial economic losses in aquaculture. Early and accurate detection of EUS is crucial for disease management, food security, and the protection of fish farmers' livelihoods. Conventional diagnostic methods, including histopathology and molecular assays, though reliable, are often time-consuming, costly, and dependent on specialized expertise. To address these challenges, Artificial Intelligence (AI) offers a promising alternative by enabling rapid, automated, and precise disease detection. This study explores the application of AI techniques, including machine learning, deep learning, and image-based recognition, in diagnosing EUS in freshwater edible fish. By analyzing fish images and associated datasets, AI models can identify characteristic disease patterns with high sensitivity and specificity. The integration of AI tools with aquaculture health monitoring not only reduces diagnostic delays but also supports the development of predictive disease surveillance systems. The findings highlight the potential of AI-driven platforms to revolutionize fish health management, minimize economic risks, and promote sustainable aquaculture. This research underscores the role of interdisciplinary approaches in advancing precision aquaculture and safeguarding aquatic resources.

KEYWORDS

Epizootic ulcerative syndrome, Freshwater edible fish, Artificial intelligence, Machine learning, Deep learning, Image recognition, Aquaculture, Fish disease detection, Precision aquaculture, Predictive surveillance.

CHAPTER 1

Epizootic Ulcerative Syndrome (EUS)

INTRODUCTION

Epizootic Ulcerative Syndrome (EUS) was delineated at a DFID Regional Seminar in Bangkok in 1994 as "a seasonal epizootic condition affecting freshwater and estuarine warm water fish, characterized by a complex infectious etiology, the presence of invasive Aphanomyces infection, and necrotizing ulcerative lesions that typically elicit a granulomatous response" (Roberts *et al.*, 1994a). Nevertheless, research conducted since then, as outlined in the aetiology and epidemiology portions of this book, indicates that a complicated aetiology is not invariably present in all cases. The case definition for EUS may now have a reference to the specific fungal infection *Aphanomyces invadans*. Given these advancements, EUS may be seen as transcending the classification of a syndrome; yet, the term "epizootic ulcerative syndrome" is widely recognized among fish health professionals and will persist in usage for this ebook.

A prior review, released by AAHRI and NACA in 1992, consolidated a significant amount of literature on the topic from national and international articles, reports, and conference proceedings. This article aims to provide practical applications to aid fish health professionals in diagnosing and managing EUS. The annex section contains extensive information on fungal and viral isolation and identification, a framework for outbreak investigations, and EUS reporting datasheets. It is important to note that other separate ulcerative fish diseases exist, and a valid EUS diagnosis can only be established through histological confirmation of the specific characteristics outlined on page 31. Consequently, it is anticipated that this manual will motivate fish health professionals to explore alternative causes of ulcerative disease. For more than 25 years, outbreaks of an ulcerative disease, histologically characterized by mycotic granulomas, have impacted freshwater and estuarine fish across much of Asia and Australia. The condition is referred to by many names: Mycotic Granulomatosis (MG) in Japan, Red Spot Disease (RSD) in Australia, and Epizootic Ulcerative Syndrome (EUS) in Southeast and South Asia. MG, RSD, and EUS were previously characterized as separate entities; however, recent research indicates that the same pathogenic Aphanomyces fungus is implicated in all instances (refer to the Aetiology section on "Fungi"). Consequently, a comprehensive history of EUS necessitates the inclusion of outbreaks in Japan and Australia.

Podeti Koteshwar Rao

Mycotic Granulomatosis (MG)

The initial documentation of an EUS-like condition occurred in the summer of 1971, involving farmed ayu (Plecoglossus altivelis) in Oita Prefecture, Japan (Egusa and Masuda, 1971). The defining lesion, a granulomatous reaction to invasive hyphae, was identified, leading to the designation of the condition as mycotic granulomatosis (Miyazaki and Egusa, 1972). It swiftly disseminated to multiple Prefectures and impacted various fish species, primarily cultured ayu and goldfish (Carassius carassius auratus), as well as wild Formosan snakehead (Channa maculata), crucian carp (Carassius auratus), bluegill (Lepomis macrochirus), and grey mullet (Mugil cephalus) (Miyazaki and Egusa, 1972; 1973a; b; c). Notably, common carp (Cyprinus carpio) remained unaffected. Hatai *et al.* (1977) identified the invasive Oomycete fungus from afflicted fish and later designated it Aphanomyces piscicida (Hatai, 1980). A. Piscicida is now recognized as conspecific with the EUS pathogen, Aphanomyces invadans (Lilley *et al.*, 1997a; b). Despite the absence of significant MG epizootics in Japan since 1973, outbreaks have persisted intermittently. Hatai *et al.* (1994) recently documented a comparable disease in imported ornamental dwarf gourami (Colisa lalia) from Singapore, which was also shown to be associated with the same Aphanomyces pathogen (Lilley *et al.*, 1997a).

In 1972, outbreaks of a cutaneous ulcerative illness known as Red Spot Disease (RSD) impacted estuarine fish, especially grey mullet (Mugil cephalus), in Queensland, Australia (McKenzie and Hall, 1976). The illness subsequently advanced to impact freshwater and estuarine fish in coastal rivers of New South Wales (Callinan *et al.*, 1989), Northern Territory (Pearce, 1990), and Western Australia. Fraser *et al.* (1992) isolated an Aphanomyces fungus from afflicted fish, demonstrating its ability to induce the disease in fish by bath challenges, contingent upon the artificial abrasion of the experimental fish's skin (Callinan, 1994b). Consequently, an additional component was deemed to be implicated in the illness process. Virgona (1992) demonstrated that RSD outbreaks in estuarine fish in the Clarence River, NSW, were correlated with reduced catchment rainfall, as noted by Callinan *et al.* (1995a),who connected this to runoff originating from acid sulfate soils. Ultrastructural analysis of fish gills and skin revealed that low pH and increased levels of monomeric aluminium, indicative of estuary acidification, cause substantial lesions in fish (Sammut *et al.*, 1996). In aquarium experiments, RSD was subsequently triggered in fish subjected to sublethal levels of artificially acidified water (at pH 3 and pH 5) and pathogenic Aphanomyces spores, even with modest quantities of monomeric aluminium (Callinan *et al.*, 1996; Callinan, 1997). Similar to the pathogenic RSD-Aphanomyces has been identified as the same species as the EUS pathogen, *A. invadans* (Callinan *et al.*, 1995a; Lilley *et al.*). Subsequent to the outbreaks of MG and RSD, a progressive westward dissemination across Asia occurred of a syndrome linked to cutaneous ulceration, resulting in significant mortalities among several freshwater and estuarine fish species. The condition was designated as Epizootic Ulcerative Syndrome (EUS) in 1986 during the Consultation of Experts on Ulcerative Fish Diseases in Bangkok (FAO, 1986). EUS outbreaks have been documented in 18 countries within the Asia-Pacific area (Fig. **1**), but not all have been definitively verified as EUS following the outbreak of an ulcerative disease, presumed to be EUS, that transpired in the rivers of southern Papua New Guinea between 1975 and 1976 (Haines, 1983). In 1982-1983, there were elevated mortality rates in gudgeon (Ophieleotris aporos and Oxyeleotris heterodon) from upland regions and mullet from estuaries in northern Papua New Guinea (Coates *et al*., 1989). Introduced tilapia (Oreochromis mossambicus) is prevalent in certain regions and has demonstrated resistance. Roberts *et al.* (1986) subsequently studied the preserved afflicted fish and confirmed the findings.

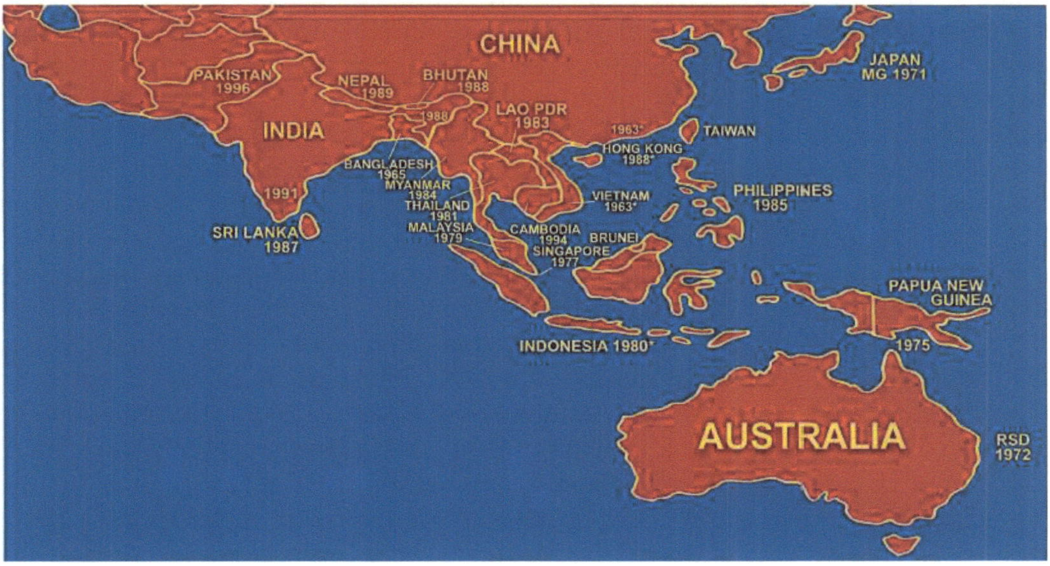

Fig. (1). A map illustrating the distribution of EUS in the asia-pacific area. there is uncertainty about the dates that signify the onset of the initial significant outbreak.

In 1980, outbreaks of an epizootic hemorrhagic syndrome transpired in Java, Indonesia, especially impacting cultured cyprinid and clariid fish, although the identification of this as EUS remains disputed (Roberts *et al*., 1986). Ulcerated snakeheads and catfish have been later documented in the Indonesian regions of Sumatra, Sulawesi, and Kalimantan (Widagdo, 1990). Invasive hyphae have been detected in sand gobies (Oxyeleotris marmoratus) from eastern Kalimantan (Rukyani, 1994), and D. Bastiawan (personal communication) separated.

Roberts *et al.* (1986) examined unverified reports of ulcerated walking catfish (*Clarias batrachus*) in Singapore in 1977 and similar incidents thereafter. Notwithstanding Singapore's position as a hub for trading in EUS-susceptible ornamental fish, there are no documented instances of significant EUS-related outbreaks. Despite indications of elevated fish mortality rates in southern peninsular Malaysia in 1979 (Shariff and Law, 1980; Roberts *et al.*, 1986), the inaugural documented instances of typical EUS outbreaks occurred in December 1980 among rice-field fishes in northern Malaysia (Jothy, 1981). These have occurred annually since then, although to a diminished degree (Shariff and Saidin, 1994). The primary species impacted include snakeskin gourami (Trichogaster pectoralis), striped snakehead (*Channa striata*), and climbing perch (Anabas testudineus. Substantial, thoroughly documented epizootics have transpired annually in Thailand since 1981 (Ulcerative Fish Disease Committee, 1983; Chulalongkorn University, 1983, 1985, 1987). The second (1982-3) and third (1983-4) epidemics were more catastrophic, impacting the extensive aquaculture systems of central Thailand and wild fish in natural watercourses. Significant mortalities occurred among cultivated snakeheads and rice-field fish. The initial outbreaks commenced at the conclusion of the rainy season (September) and continued throughout the cool season until March. Outbreaks are now typically confined to the colder months of December and January. In December 1996, EUS was seen in the northeastern, central, and southern regions of Myanmar, Lao PDR, and Cambodia, and was initially reported in significant outbreaks in 1983 or 1984 (Lilley *et al.*, 1992). Later epizootics were less widespread (*e.g.*, EUS impacted 35 Burmese townships in 1984-85 and 11 townships in 1989-90: Soe, 1990). The significance of vulnerable fish to rural people in these nations ensures that the impact remains considerable. In 1996, infected snakeheads from Laos were verified at AAHRI, Bangkok, as numerous reports of EUS-affected fish have emerged from Vietnam, China, and Hong Kong, although they remain unverified. The initial documentation of ulcerated snakeheads in Vietnam, indicating the probable first incidence of EUS in the nation, originated from the Mekong Delta in 1983 (Xuan, 1990). Ulcerated Labeo rohita were initially documented at the Pearl River Fisheries Institute in Guangzhou, South China, in 1982 (Lian, 1990). Clariid catfish were impacted in the same region during 1987-1988 (Lian, 1990), while Carassius auratus were apparently afflicted throughout much of Eastern China in 1989 (Guizhen, 1990). Wilson and Lo (1992) documented seasonal mortality rates of up to 70% in snakeheads (*Channa maculata*).

In December 1985, Laguna de Bay in the Philippines encountered a significant outbreak of EUS. An estimated 5-40% of snakeheads, gobies, gouramies, catfish, crucian carp, and Arius species were ulcerated, while milkfish, bighead carp, and tilapia remained undamaged (Llobrera and Gacutan, 1987). The disease persisted in disseminating to a minimum of 11 other provinces, impacting wild fish in lakes, rice fields, marshes, and farmed fish in ponds (Bondad-Reantaso *et al.*, 1994). Mullet, goatfish (Upeneus bensai), croaker (Johnius spp.), and *Psettodes* spp. were affected, and an outbreak occurred in 1990, identified as EUS through histological analysis (Reantaso, 1991; S. Chinabut, unpublished). The presence of EUS in these brackish and marine species elucidated the potential mechanisms of its transmission between the islands. The intensity of outbreaks has diminished since 1993. Numerous *A. invadans* isolates were obtained from EUS-affected fish in the A significant epidemic of EUS in freshwater and estuarine fish in western Sri Lanka transpired in December 1987, preceding any outbreaks on the mainland of the subcontinent (Costa and Wijeyaratne, 1989). The disease is believed to have been introduced from Southeast Asia by exports of sick fish, potentially ornamental angel fish (Pterophyllum scalare), some of which had ulcers and experienced elevated mortality rates (Balasuriya, 1994). Snakeheads exhibiting substantial necrotic ulcers were the most prominent indicator of the disease; nonetheless, tilapia, the primary commercial species, remained unaffected. In 1996, EUS was allegedly still operational in Batticaloa lagoon (*P. Vinobaba*).

In the last decade, EUS has significantly impacted fisheries across mainland South Asia, resulting in losses in vital capture fisheries regions and undermining trust in an aquaculture sector that is still nascent. The ailment was initially documented in Chandpur district, Bangladesh, in February 1988. The initial outbreak persisted for 13 months, during which it proliferated swiftly across the nation, allegedly facilitated by the flood of September 1988 (Barua, 1994). Ulceration was noted in numerous natural species, primarily snakeheads, Puntius, Clarias, Mystus, and Mastacembelus. Cultured Indian major carp were impacted; however, significant mortalities from the disease were likely confined to fingerlings (Roberts *et al.*, 1989). The prevalence of EUS subsequently decreased; nevertheless, indications indicate that since 1995, the severity of outbreaks in Bangladesh has been escalating (G.U. Ahmed, unpublished study). In January 1993, A. Invasive cultures were separated from cultivated Indian major carp (Labeo rohita) in Northwest Bangladesh and from wild fish in the fertile floodplain region of Northeast Bangladesh. Outbreaks of EUS in India have been thoroughly examined (Zoological Society of Assam, 1988; Jhingran and Das, 1990; National Workshop on Ulcerative Disease Syndrome in Fish, 1990; Kumar *et al.*, 1991; ICSF, 1992; Das and Das, 1993; Mohan and Shankar, 1994). The northeastern Indian states were the initial regions to declare losses in May 1988. The disease seemed to disseminate *via* rivers, lakes, and paddy fields to numerous states, impacting several Indian big carp farms as well. EUS significantly affected fish populations in the low salinity regions of the abundant brackishwater fishery of Chilka Lake, Orissa, in November 1990 (Raman, 1992), as well as in the reservoirs and backwaters of Kerala in June 1991 (Sanjeevaghosh, 1991). Aphanomyces isolates aligned with *A. Invadans* have been. In 1989, Bhutan and the eastern Terai of Nepal were initially impacted, and by 1993, EUS had disseminated to Himalayan valley regions such as Pokhara and Kathmandu, affecting cold water species, including Tor spp. (Phillips, 1989; Shresta, 1994). It is estimated that 20-30% of Nepalese pond fish production, around 3000 metric tons, is lost annually due to EUS (Pantha). The most recent country impacted by EUS was Pakistan, where the disease was confirmed in snakeheads from Punjab Province in April 1996 and in Cirrhinus mrigal from Sindh Province in January 1998 (DFID, 1998). The blotched snakehead or mud murrel (Channa punctata) was the most often impacted species, with Puntius spp., Labeo rohita, and Cirrhinus reba also reported as affected (N. Akhthar, pers. comm.). Approximately 20% of farms in the Sialkot Division of Punjab were impacted, with a greater frequency observed in ponds inundated by the 1996 flooding (AAHRI, ACIAR, IoA, and NACA, 1997). Documented losses in Punjab have been minimal, likely attributable to the widespread utilization of tube-well water for aquaculture and increased salinities (AAHRI, ACIAR, IoA, and NACA, 1997). However, EUS has become firmly established in certain regions of the Indus River, and its seemingly swift proliferation throughout the nation (DFID, 1998) raises concerns regarding potentially significant future repercussions. Noga (1994) proposed that Ulcerative Mycosis (UM) in coastal fish populations of the western Atlantic may be part of the same syndrome as Epizootic Ulcerative Syndrome (EUS), due to the similarities in clinical and pathological characteristics of both diseases and the predominant recovery of Aphanomyces fungi from UM-affected fish (Dykstra *et al.*, 1986). Nonetheless, fish exposed to these *Aphanomyces* isolates have not exhibited lesions characteristic of UM (Noga, 1993; Lilley and Roberts, 1997). Fish exhibit UM when lesion material serves as an inoculum,

UM was initially documented in April 1984 in menhaden (Brevoortia tyrannus) in the Pamlico River, North Carolina, with a significant mortality event reported in November of the same year (Noga and Dykstra, 1986). Epidemics of analogous diseases were subsequently identified in estuaries along the eastern shore of the USA, from Connecticut (Noga, 1993) to Florida (McGarey *et al.*, 1990). Still, it remains questionable whether these were initial occurrences or indicated a dissemination of the disease. The prevalence in these species was significantly lower than that in menhaden (Levine *et al.*, 1990a;Noga *et al.*, 1991) in the Pamlico River. A greater percentage of age-0 menhaden were found to be impacted compared to age-1 fish (Levine *et al.*, 1990b). Levine *et al.* (1990b) presented evidence indicating that particular low salinity areas within the Tar-Pamlico estuary contained elevated levels of diseased fish, while Noga (1993) noted that the most severe outbreaks in the Pamlico River correlated with years of outbreaks persisting, with menhaden infection rates reaching 100% (Levine *et al.*, 1990b). Munday (1985) documented the occurrence of severely ulcerated red cod

(Pseudophycis bachus) in the River Tamar in Launceston, Tasmania, in November 1980 and 1981. Despite the identification of many germs and parasites in the fish, pollution was deemed the primary cause of the disease. Munday (pers. comm.) currently posits that ulcer illness was identical to EUS, albeit occurring in higher salinity water; nevertheless, he notes that since the enhancement of Launceston's sewerage system, the condition is no longer documented.

Impacted Species

Over 100 fish species have been documented as affected by EUS (Lilley *et al*, 1992), yet only a limited number of reports have been substantiated through the identification of mycotic granulomas in histological sections or by isolating the pathogenic Aphanomyces fungus from tissues beneath ulcers. Species identified as unaffected by EUS epidemics comprise Chinese main carps, tilapias, and milkfish (Chanos chanos). Hatai (1994) conducted experimental injections of hyphae Aphanomyces into catfish (Parasilurus asotus), loach (Misgurnus anguillicausatus), and eel (Anguilla japonica) and determined these species to be resistant to infection. Wada *et al.* (1996) and Shariffpour (1997) experimentally inoculated common carp (Cyprinus carpio) with zoospores of Aphanomyces from MG and EUS outbreaks, respectively, and established that fungal proliferation was inhibited by a robust host response. Certain authors have noted that the species most adversely impacted during natural outbreaks are typically benthic organisms (Llobrera and Gacutan, 1987; Chondar and Rao, 1996) or those with aerial respiration capabilities (Roberts *et al.*, 1994b). No specific size group of snakeheads is particularly vulnerable, as affected fish vary from 40g to 900g (Cruz-Lacierda and Shariff, 1995). Nonetheless, size or age may be pertinent in other species. Indian major carp experience significant mortality rates as fingerlings (Roberts *et al.*, 1989), larger specimens, despite exhibiting ulcerations, are not documented to perish in substantial quantities (AAHRI, ACIAR, IoA, and NACA, 1997).

This indicates the potential for additional dissemination of the disease to these regions. It is important to acknowledge that the ideal temperatures for *in vitro* vegetative growth of A. Invadans thrive within a temperature range of 20-30°C (Fraser *et al.*, 1992; Lilley and Roberts, 1997), which likely accounts for the fact that natural outbreaks have thus far been confined to latitudes between 35°N and 35°S. Experimental injection trials with native European and American fish species have demonstrated that the pathogenic fungus, *Aphanomyces invadans*, can induce lesions in rainbow trout at 18°C (Thompson *et al.*, in press) but exhibits reduced infectivity in stickleback (Gasterosteus aculeatus) and roach (Rutilus rutilus) at 11-16 °C. The most significant effect of EUS has likely been on small-scale, mixed-species fishing and aquaculture in rice fields and rural waterways. Approximately 250 million families in the Asia-Pacific area rely on rice as their primary crop, and the incidental fish harvested from these fields constitutes a significant component of their diet (Macintosh, 1986). The primary months for harvesting rice paddy fish are from September to February, coinciding with the peak incidence of ulcerative diseases. Under these circumstances, any estimate of the financial cost of EUS may undervalue the complete burden of the disease on these communities.

These statistics, however, do not consider secondary socio-economic costs resulting from the market rejection of harvested ulcerated fish, or, in certain instances, even uninjured fish. During the 1980s, several localities experienced a pervasive albeit baseless apprehension of illness transmission to customers, resulting in a significant decline in market demand for all food fish. In the Philippines, the average daily income of fishers, roughly US$4, decreased to US$1.50 during disease outbreaks in Laguna de Bay due to the rejection of contaminated fish (ADB/NACA, 1991). Bangladesh had significant losses due to EUS in 1988 and 1989, and widespread local media coverage of the disease exacerbated public anxiety around health hazards associated with fish intake, leading to early price declines of up to 75% and substantial losses for fish traders. Nepal lacks marine fish resources and hence depends significantly on EUS-susceptible species. Reports indicated that 15-20% of overall fish productivity in Nepal was lost during the initial outbreaks of EUS (ADB/NACA, 1991). The emergence of EUS in cultured big carp fingerlings has raised concerns about its potentially debilitating impact on the growth of carp aquaculture in the subcontinent. Bhaumik *et al.* (1991) indicated that 73% of the culture ponds in West Bengal were impacted at that time, with the majority experiencing a loss of 30-40% of their stock. The EUS pandemic as reported by the ICSF (1992) has illustrated to national authorities the capacity of fish disease to inflict significant financial losses; consequently, a beneficial outcome of EUS has been the augmented funding directed towards fish disease research and before the initial EUS outbreaks, most countries in the region had not encountered a fish disease epizootic of such magnitude, leading to considerable local apprehension regarding the implications of consuming diseased fish or utilizing contaminated waters for domestic or agricultural purposes. The simultaneous deaths of ducks, cattle, and other animals were attributed to EUS. Nonetheless, no scientific data indicates that the sickness itself causes any illness in humans or animals. Rahman *et al.* (1988) could not elicit any illness signs in ducks that were fed EUS-infected fish or injected with Aeromonas hydrophila cells. Consequently, it is essential to implement public educational initiatives to alleviate the inherent concerns of farmers, fishers, and consumers regarding the broader implications of EUS. It is imperative to emphasize the necessity of adhering to proper hygiene habits. Specifically, deceased fish should not be harvested for sale or consumption, not because of ulcerative disease per se, but because germs or poisons found in decomposing, EUS-affected fish. The indiscriminate application of chemotherapeutants for the treatment of EUS or other ailments in intensive culture systems raises significant public health concerns. Chloramphenicol, for example, is utilized in the treatment of typhoid in humans, and there exists a concern that the accumulation of bacterial resistance in treated fish (Poonsuk *et al.*, 1983) may be transmitted to people. Farmers are most concerned about the potential for serious allergic reactions among farm workers exposed to the medicine. There exists a risk that customers may encounter drug residues in commercially available fish that were hastily collected prior to the completion of the specified withdrawal time. The prevalence of EUS has underscored the need to establish appropriate rules and regulations to safeguard farmers and consumers from the indiscriminate application of chemo-therapeutants, despite these broader issues impacting aquaculture.

Fish afflicted with skin ulcers are susceptible to opportunistic microorganisms, and in chronic instances, determining the origin of the original lesion can be challenging. Due to the extensive geographical range and varied habitats of EUS-affected fish, a notably broad array of microbial agents has been isolated from these species. Certain agents may substantially contribute to a disease complex during a specific outbreak; nonetheless, it is crucial to differentiate them from the factor or factors that are fundamental to all EUS epidemics. This text provides a description of fungi, viruses, bacteria, and parasites linked with EUS lesions, along with commentary on recent research. Recent research has established that a singular species of Aphanomyces "fungus" is an essential causative agent of EUS, as it is present in all outbreaks. In certain outbreaks (*e.g.*, in Australian estuaries), it may be the sole biological *Aphanomyces* species recognized as contributors to the etiology of EUS in Southeast Asia since the early outbreaks in Thailand. Limsuwan and Chinabut (1983) characterized a "severe chronic granulomatous mycosis" in the histological sections of afflicted fish. The prevalence of saprophytic fungal pollutants on EUS lesion surfaces resulted in the discovery of Achlya and Saprolegnia species. These were promptly identified as secondary agents (Tonguthai, 1985).

As described in the History section, prior to the initial emergence of EUS in Southeast Asia, the pathogenic Aphanomyces piscicida was identified from MG-affected fish in Japan (Hatai *et al.*, 1977), but MG had not yet been acknowledged as synonymous with EUS. The *Aphanomyces* fungus was subsequently isolated from RSD outbreaks in Australia in 1989 (Fraser *et al.*, 1992) and, independently, from EUS outbreaks in Thailand during 1991-1992 (Roberts *et al.*, 1993). These isolates demonstrated the ability to induce characteristic EUS lesions when injected subdermally into susceptible fish. All pathogenic isolates of MG, RSD, and EUS were demonstrated to be slow-growing and thermolabile in culture. Comparable isolates have also been acquired from the Philippines, Indonesia, Bangladesh (Lilley and Roberts, 1997), and India (I. Karunasagar, pers. comm.). Recent comparisons of pathogenic Aphanomyces cultures from various countries have demonstrated, through protein banding profiles (Callinan *et al.*, 1995b; Lilley *et al.*, 1997b), growth characteristics (Lilley and Roberts, 1997), and chemical susceptibility (Lilley and Inglis, 1997), that they belong to the same species. Genetic fingerprinting methods have demonstrated that the different isolates are highly similar at the genetic level (Lilley *et al.*, 1997a). This evidence indicates that the isolates are not permanent inhabitants of each place, as would be anticipated for an opportunistic fungus. Rather, they constitute a singular fungal strain that has proliferated throughout most of Asia and Australia.

The pathogenic Aphanomyces has been referred to as Aphanomyces piscicida (Hatai, 1980), Aphanomyces invaderis (Willoughby *et al.*, 1995), and ERA (EUS-related *Aphanomyces sp.* Lumanlan-Mayo *et al.*, 1997). Since the isolates in each instance have been demonstrated to be conspecific, a single species name is necessary to characterize all these isolates. According to the International Code of Botanical Nomenclature (ICBN), "invadans" is the sole legitimate taxon name. *A. invadans* is recognized for its rapid growth in culture at temperatures ranging from 26 to 30°C (Hatai and Egusa, 1978; Fraser *et al.*, 1992; Lilley and Roberts, 1997) has demonstrated growth in snakehead muscle tissue at temperatures between 19 and 31°C (Chinabut *et al.*, 1995). Further studies have demonstrated that snakeheads can recuperate from *A. invadans* infection that occurs at elevated temperatures (26, 30°C), although they fail to inhibit fungal invasion, ultimately resulting in disease manifestation at reduced temperatures (19°C) (Chinabut *et al.*, 1995). The humoral and cellular immune responses of fish are known to be inhibited at low temperatures (Avtalion *et al.*, 1980; Bly and Clem, 1991), which may elucidate the increased mortalities from EUS during periods of low water temperatures. Both naturally and experimentally infected snakeheads have demonstrated an antibody response against Aphanomyces. The cellular macrophage response is deemed crucial for fish to combat infection. A compilation of the diverse published accounts detailing the attributes of the invasions from EUS, MG, and RSD epidemics is presented.

CONCLUSION

The Epizootic Ulcerative Syndrome (EUS) presents a considerable threat to global aquaculture, impacting numerous fish species. The effective management of EUS depends on a synthesis of precise diagnostic methods, preventive actions, and control strategies. Advanced diagnostic instruments, such as molecular methods like PCR, along with histopathological and histochemical analyses, have improved the early detection and surveillance of the disease. Preventive approaches, including enhancing water quality, regulating environmental stressors, and implementing biosecurity standards, are crucial in reducing epidemics. Molecular diagnostics and environmental monitoring are crucial for prompt response. Preventive interventions, such as water quality control, biosecurity measures, and stress alleviation, are essential in diminishing vulnerability to the disease. Control strategies, encompassing antifungal therapies, vaccine development, and the amalgamation of conventional and contemporary management techniques, are crucial for alleviating the effects of EUS. Ongoing research and collaboration among stakeholders, including fish farmers, researchers, and policymakers, are essential for promoting sustainable aquaculture methods. By implementing comprehensive, proactive measures, the aquaculture sector can effectively address EUS, safeguard fish health, minimize economic losses, and foster environmental sustainability. Continued investigation into disease processes, host-pathogen interactions, and innovative therapeutic strategies will facilitate enduring solutions to EUS. Epizootic Ulcerative Syndrome (EUS) remains a significant threat to aquaculture, causing substantial economic losses and impacting aquatic biodiversity. Effective management of EUS necessitates a comprehensive approach that includes advanced diagnostic methods, strong prevention strategies, and dependable control measures.

AI and ML-driven preventive techniques represent a significant transformation in aquaculture disease management. Predictive modeling facilitates the early detection of outbreak hazards, whereas optimization methods guarantee sustainable and efficient agricultural operations. Decision support systems facilitate equitable access to sophisticated technology, enabling farmers and aquaculture managers to proactively address EUS hazards. These innovations collectively improve resilience, productivity, and sustainability in global aquaculture systems. The management of Epizootic Ulcerative Syndrome (EUS) in aquaculture is on the verge of transformation through the implementation of Artificial Intelligence (AI) and Machine Learning (ML). These technologies provide robust solutions for the issues associated with disease detection, prevention, and control, hence maintaining the sustainability of aquaculture activities.

Revolutionizing EUS Management with AI and ML

Artificial Intelligence and Machine Learning facilitate a transformative change in Endoscopic Ultrasound management by implementing:

Improved Diagnostic Precision

AI-driven picture recognition and predictive models provide swift and dependable disease identification, exceeding conventional techniques in both speed and accuracy. Data-Driven Prevention, predictive analytics, and decision-support systems provide proactive interventions utilizing environmental and historical data, hence mitigating outbreak risks. Efficient Control Strategies: AI-driven treatment recommendation systems, together with continuous monitoring, enhance interventions while reducing resource consumption and environmental impact. These inventions enhance fish health and farm productivity while also addressing global issues such as food security and sustainable aquaculture. The Significance of Technology Adoption: To harness the potential of AI and ML, it is essential to overcome adoption barriers and cultivate a collaborative ecosystem. Empowering small-scale farmers through accessible, cost-effective technologies tailored to their needs can democratize the advantages of AI, facilitating widespread adoption. Advancing Sustainability: AI technologies enhance resource efficiency, reduce waste, and lessen dependence on harmful practices, harmonizing aquaculture with environmental conservation objectives. By incorporating real-time monitoring, early warning systems, and sophisticated analytics, aquaculture operations can more effectively respond to challenges such as climate change and disease outbreaks and enhance resilience. The integration of AI and ML is essential for the future of sustainable aquaculture. By embracing these technologies, governments, researchers, technology developers, and farmers can advance the aquaculture industry towards a future where EUS and other illnesses are effectively managed, safeguarding the health of aquatic ecosystems and the lives of millions reliant on them. Artificial Intelligence and Machine Learning provide the means to harmonize productivity with sustainability, facilitating a robust and thriving aquaculture industry.

CHAPTER 2

Hematological Analysis

INTRODUCTION

Blood serves as an effective signal for assessing the health of an organism (Joshi *et al.*, 2002a). There is a growing recognition among fish pathologists regarding the utilization of hematological and biochemical parameters as diagnostic instruments for assessing the physiological health status of fish. Joshi *et al.* (2002b) and Chagas and Val (2003) indicated that hematological parameters serve as a pathological indicator for the entire organism and are crucial for evaluating the physiological condition of fish affected by pathogens and for diagnosing their functional status. The initial study on the hematology of Indian fish was likely published by Dhar in 1948. This was the first study on the shape of erythrocytes and leukocytes, clotting time, *etc.*, in the air-breathing fish Ophiocephalus punctatus. Das (1958) reported the haematological data from three Indian main carps: Labeo rohita, Catla catla, and Cirrhinus mrigala. Houston and DeWilde (1968) proposed in their study on Salmogiardnerii, *Cyprinus carpio*, and Salvelinus fontinalis (brook trout) that hematocrit values should serve as a general indicator of hematological state in regular examinations.

Hematological and biochemical assays are not extensively utilized for diagnosing fish diseases; nonetheless, these tests could serve as effective diagnostic tools for monitoring physiological and pathological alterations in fish. Various factors, including season (Mahajan *et al.*, 1979b), water contamination (Witeska, 1998), physicochemical parameters of water stress (Joshi, 1982; Nikoo *et al.*, 2010), age and sex (Svetina *et al.*, 2002), and fish species, influence haematological parameters (Nikoo *et al.*, 2012).

Blood is a vital fluid connective tissue of the body. It constitutes a vascular tissue. Blood forms the organism's internal milieu and performs numerous functions. It executes the essential processes of metabolism and respiration. Blood necessitates all conditions requisite for the provision of oxygen (O2), vital for tissue oxidation. It functions as a conduit for the elimination of metabolic wastes and metabolites from a system through excretion. The quality and quantity of blood components will aid in assessing the physiological condition of the animal. Hematological studies in fish are typically conducted for toxicological research, environmental monitoring, and as potential indicators of physiological or pathological changes in fishery management and disease investigations. The data also aid in assessing physiological responses to specific stressors and in establishing hematological guidelines for different animals (Kawamoto 1929; Khan and Siddiqui, 1969).

The escalation of urbanization, industrialization, population growth, and humanity's insatiable greed to exploit natural resources has engendered a significant threat to all forms of life, culminating in a global crisis. Substantial quantities of industrial waste, pesticides, and other organic chemicals and fertilizers utilized in agriculture readily infiltrate water bodies *via* runoff, leading to water pollution. Aquatic pollution is particularly concerning, as all forms of life rely on water. Natural ecosystems are contaminated by industrial waste and other anthropogenic activities among all forms of water contaminants (Velez and Montoro 1998; Conacher *et al.*, 1993). Industrial wastes represent the primary cause of metal pollution in natural water (Livingstone, 2001). Hematological indicators are critical criteria for assessing the physiological health of fish. Their alterations are contingent upon fish species, age, the reproductive cycle of spawners, and diseases (Luskova 1997; Golovina Trombicky 1989; Golovina, 1996). Freshwater fish exhibit a close relationship with their environment, rendering them highly vulnerable to physical and chemical fluctuations, which may manifest in their blood constituents.

Blood is acknowledged as a potential indicator of fish responses to water quality and can be utilized to evaluate the impact of environmental toxins, as outlined by Hesser (1960) in procedures for routine fish hematology. Larson and Snieszko (1961) evaluated different techniques for measuring haemoglobin in trout blood. Bouck and Ball (1966) asserted that haematology might serve as a valuable instrument for assessing the stress levels in fish due to water pollution. Gelineo (1969) examined the concentration of hemoglobin in 53 species of freshwater and marine fish. Blaxhall and Daisely (1973) delineated standard haematological techniques for analyzing fish blood, encompassing haemoglobin quantification, Packed Cell Volume (PCV), erythrocyte enumeration, Erythrocyte Sedimentation Rate (ESR), Total Leukocyte Count (TLC), Differential Leukocyte Counts (DLC), and cytochemical staining. This is a description of stained blood cells, along with the range and mean values for these tests on brown trout (Salmo trutta) as reported by Hussein *et al.* (1974), who conducted a hematological investigation on Anguilla vulgaris and Mugil cephalus. Seasonal fluctuations in erythrocyte count, haematocrit values, and haemoglobin content have been noted for both species, with elevated levels in summer and diminished levels in winter. Denten and Yousuf (1975) documented seasonal alterations in the haematology of Salmo gairdneri. Clark *et al.* (1979) investigated physiological stress induced by environmental factors in largemouth bass, Micropterus salmonids, and found that Haematocrit (Ht), Haemoglobin (Hb), and total plasma protein exhibited a positive correlation with fish length; Hb and Ht were positively correlated with fish age, whereas mean corpuscular haemoglobin showed a negative correlation with fish age. Haemoglobin and packed cell volume were correlated with erythrocyte counts, as noted by Siddiqui and Nasim (1979), who observed that *Cirrhina mrigala* had higher haemoglobin and erythrocyte concentrations in males compared to females.

Wedemeyer *et al.* (1983) examined the physiological stress response in Oncorhynchus kisutch and determined that hematocrit levels were a sensitive predictor of physiological stress caused by population density and temperature fluctuations. Kumar *et al.* (1984) investigated the haematological alterations in the cold water fish Schizothorax plagiostomus (Heckel), which were naturally infected with metacercariae of Diplostomium tertare. The study revealed a decrease in total erythrocyte count, Packed Cell Volume (PCV), Hemoglobin (Hb), and Total Leukocyte Count (TLC) compared to controls, indicating pollution. Murray (1984) correlated haematological traits with the sex of fish and the seasonal period. Ralio *et al.* (1985) indicated that fish blood parameters, including erythrocyte and leucocyte counts, hemoglobin, hematocrit, and leucocyte differential counts, are significantly influenced by incidental factors such as physical and environmental stress from water contaminants. Rao *et al.* (1989) noted that active, fast-moving fish (Scomberomorus guttatus and Rastrelliger kanagurata) exhibited elevated erythrocyte parameters to accommodate their high metabolic rates compared to sluggish predatory fish (Arius maculates) and bottom detritus feeders (Liza parsia).

The leukocyte characteristics were unrelated to the fish's activity and habitat. Thakur and Pandey (1990) investigated the impact of BHC on *Clarias batrachus* and determined that lymphocytosis, neutropenia, and eosinopenia were associated with BHC toxicity. Hoglund *et al.* (1992) investigated the haematological changes in a population of Anguilla anguilla naturally infected with Anguillicola crassus throughout the Swedish Baltic coast, specifically in an area receiving heated cooling water from a nuclear power plant. Infection led to decreased lymphocyte counts and elevated granulocyte counts, suggesting a humoral and cellular immune response that influenced hematological parameters in Oreochromis aureu*s* (Allen, 1993).

The findings indicated that parameters of Aurens seem to resemble those of *O. Niloticus* and *M. mossambicus*. Hemoglobin concentration was elevated in O. aureus and subsequently in other species. Rauthan and Grover (1994) noted that blood parameters are influenced by both intrinsic and external variables. Research on the hematological parameters of Barilius bendelisis across various seasons indicated an increase in total erythrocyte count, hemoglobin, packed cell volume, and blood glucose levels during summer, while a decline in all parameters was noted in winter, characterized by significantly low ambient temperatures (Hrubec *et al*., 1997).

These data reveal a seasonal fluctuation in the blood of male and female Tinca tinca. Sahoo and Mukherjee (1999) determined the normal ranges for clinically significant haematological parameters in laboratory-reared Rohu (Labeo rohita) fingerlings. Several researchers have examined the impacts of stressors such as infections and pharmaceuticals.

Gabriel *et al*. (2004) investigated the effects of sex, acclimation source (pond and wild), and health state on various blood parameters of *Clarias gariepinus*. The findings of this study indicate that sex, fish source, and acclimation duration exert varying degrees of influence on the blood parameters of African catfish (*Clarias gariepinus*). Jawad *et al*. (2004) examined the correlation between haematocrit, body length, sex, and reproductive condition in the Indian Tenalosa ilisha, revealing that male fish had greater haematocrit levels than females.

Rehulka and Adamec (2004) investigate female rainbow trout to determine reference hematological values for red blood cell counts, hematocrit levels, and hemoglobin concentration. Tierney *et al*. (2004) investigated the differential leukocyte count of four teleost species: Coho salmon (Oncorhynchus kisutch), Pacific herring (Clupea pallasi), brook stickleback (Culaea inconstans), and feathered minnow (Pimephales promelas), and found that the relative leukocyte count significantly responds to variations in water quality. Arnold (2005) established standardized hematological methods and reference intervals for cartilaginous fishes, including sharks, skates, and rays. De-Pedro *et al*. (2005) investigated daily and seasonal variations in hematological and biochemical blood parameters in tench, Tinca tinca. Elahee and Bhagwant (2005) investigated gill histology and hematological indicators in three tropical marine fish species: Scarus ghobban, Epinephelus merra, and Siganus sutor, from the apparently contaminated lagoon of Bain des Dames, Mauritius. Gbore *et al*. (2006) investigated the impact of stress from handling and transportation on hematology and plasma biochemistry in the fingerlings of two fish species, Tilapia zilli and C. gariepinus, concluding that fingerlings are generally vulnerable to stress. Zilli exhibits greater susceptibility to physical stress compared to *C. garipinus*.

Gabriel *et al*. (2007) delineate the hematological attributes of Sarotherodon melanotheron from the brackish water creek of Buguma. The maximum range of characteristics was noted in thrombocytes, whilst the minimum values were detected in RBC. Significant variations between males and females were identified in hemoglobin, hematocrit, red blood cell count, and thrombocyte levels. Sahan *et al.* (2007) conducted a study in the agricultural, industrial, home, and slaughterhouse discharge areas of the Ceyhan River, revealing that leukocyte levels and neutrophil proportions in fish blood were elevated due to environmental stressors. Zexia *et al.* (2007) conducted morphological research on the peripheral blood cells of the Chinese sturgeon, Acipenser sinensis. Erythrocytes and the four primary types of leukocytes, namely thrombocytes, lymphocytes, granulocytes (comprising neutrophils and eosinophils), and monocytes, were found in the peripheral blood system. Furthermore, the standard counts of erythrocytes, reticulocytes, and erythrocyte division were recorded.

Adam and Agab (2008) examined the reference values for hematological and biochemical parameters of *Clarias gariepinus*. The red blood cell count had a positive correlation with haemoglobin and a negative correlation with MCV and MCH. Bastami *et al.* (2008) conducted a study to acquire fundamental insights into the hematology and the impact of sex on various blood parameters of wild carp (*Cyprinus carpio*) spawners. The maximum haematocrit, haemoglobin concentration, red blood cell count, mean corpuscular haemoglobin, and mean corpuscular haemoglobin concentration were seen in males. The largest leukocyte differential numbers were observed in females. Latifi *et al.* (2007) conducted a cytomorphometric assessment of Koran (*Salmo letnica*) erythrocytes in natural settings.

The haemoglobin, red blood cell count, total leucocyte count, and packed cell volume were up in Suheli river fish, while the white blood cell count and erythrocyte sedimentation rate were greater in Gomti river fish. This study demonstrates that blood parameters serve as sensitive indicators of stress in fish subjected to water contaminants. Zhou *et al.* (2009) conducted a comparative analysis of hematology and serum biochemistry between farmed and wild ecotypes of the dojo loach, *Misgurnus anguillicaudatus*. The results indicated that haemoglobin levels differed significantly across the two ecotypes, but red blood cell counts were markedly larger in cultured individuals compared to their wild counterparts. The white blood cell count in farmed fish was markedly lower than that in wild fish. Sabri *et al.* (2009) investigated the effects of Henneguyosis infestation on the hematology of *Clarias gariepinus*. They have also noted a decrease in RBC count, Hb levels, and packed cell volume, while the WBC count has increased in the infected fish.

Nikolov and Boyadzieva-Doichinova (2010) investigated the red blood cell count in three carp species: *Carassius gibelio* (L.), *Alburnus alburnus* (L.), and *Scardinius erythrophthalmus* (L.). The results were compared with those of individual species and data from other freshwater fish. Akinrotimi *et al.* (2010) examined the hematological parameters of *Tilapia guineensis* from Bunguma

Creek, Nigeria. Dove *et al.* (2010) investigated hematological and serum biochemical parameters in the whale shark, *Rhincodon typus*.

Gabriel *et al.* (2011) examined the impact of acclimatization on the blood composition of *Oreochromis niloticus*. Following acclimatization to captivity, there was a significant decrease ($p<0.05$) in the levels of Haemoglobin (Hb), Packed Cell Volume (PCV), Red Blood Cells (RBC), Mean Corpuscular Haemoglobin (MCH), Mean Corpuscular Haemoglobin Concentration (MCHC), lymphocytes, and thrombocytes, whereas the values of Mean Corpuscular Volume (MCV), White Blood Cells (WBC), neutrophils, and monocytes exhibited a significant increase ($p < 0.05$). Kori-Siakpere and Oboh (2011) investigated the impact of tobacco leaf dust on the hematological parameters of African catfish, *Clarias gariepinus*. This result demonstrates a significant difference ($P<0.01$) in haematological parameters such as haematocrit, haemoglobin, total erythrocyte count, and mean erythrocyte haemoglobin concentration, indicating a proportionate reduction with an increase in toxicant concentration over the exposure time.

The hematological assessment revealed a reduction in erythrocyte production and a significantly lower hemoglobin concentration in the red blood cells of the exposed fish compared to the control, indicating an anemic state. Malathi *et al.* (2012) investigated the comparative hematology of *Channa punctatus* and *Channa striatus* (Bloch) sourced from freshwater bodies in the Cauvery delta. The results indicate that the haematological parameters exhibited little variation between the two species. The RBC count, hemoglobin, packed cell volume, mean corpuscular volume, and mean corpuscular hemoglobin concentration were higher in *Channa punctatus* and lower in *Channa striatus*. In contrast, the WBC count and mean corpuscular hemoglobin exhibited elevated levels in *Channa striatus* than in *Channa punctuatus*.

The blood of fish, closely linked to their aquatic environment, reflects internal conditions related to the external manifestations of disease, which comprise three primary components: the host (fish), the pathogen (stressor), and the environment (Fernandes and Mazon, 2003). Fish do not act as hosts for numerous pathogens; yet, these pathogens inflict significant damage and modify the physiology, histology, and hematology of the host (Reholka 1996; Hauston *et al*., 1997; Ranzani-Paiva *et al* ., 2000; McNulty *et al* ., 2003; Sebastiao *et al*., 2011).

Variations in the hematological and biochemical parameters of fish in response to diverse microbial and parasite illnesses have been assessed across several fish species. Smith and Hattingh (1978) examined the haematological parameters and blood cell indices of freshwater teleost fishes, such as *C. Carpio, S. mossambicus*, and *S. gaivdneri*. Lester and Budd (1979) noted alterations in the blood cells of coho salmon fingerlings experimentally infected with bacterial pathogens responsible for vibriosis and bacterial kidney disease. Barham (1980) examined the fluctuations in hematological parameters, such as hematocrit levels and Erythrocyte Sedimentation Rate (ESR), in rainbow trout exhibiting bacterial illnesses. Waagbo *et al.* (1988) analyzed the hematological and biochemical characteristics of Atlantic salmon, Salmo salar, affected by coldwater vibriosis (Hitra sickness). Haematological alterations linked to epizootic ulcerative syndrome have been documented for *Etroplus suratensis*. Pathiratne and Rajapakshe (1998), along with Cruz-Lacierda and Shariff (1995) regarding the snakehead *Ophiocephalus striatus* and Rehulka (1996) investigated blood parameters in common carp afflicted with spontaneous spring viremia (SVC). The blood parameters of common carp, *Cyprinus carpio*, experimentally infected with Pseudomonas fluorescens, were examined (Yildiz, 1998).

Rehulka (2002) found that hematological and biochemical alterations were linked to skin lesions induced by spontaneous infections with Aeromonas in rainbow trout (*Oncorhynchus mykiss*). He researched the fundamental properties of erythrocytes, blood plasma metabolites, and alterations in the plasma protein spectrum. Siwicki *et al.* (2003) documented alterations in many haematological and biochemical parameters in the blood of pike perch following intense culture, including changes in Hb%, PCV, and total cell counts of RBC and WBC (Martins *et al.*, 2008).

Variations in the haematology of African catfish (*Clarias gariepinus*) due to handling and transportation effects have been recorded. Adeyemo *et al.* (2009) and Zorriehzahra *et al.* (2010) conducted studies on blood parameters, namely. Total RBC and WBC counts, PCV, Hb%, blood cell indices MCH and MCHC, as well as blood serum parameters including total proteins and enzymes, in rainbow trout fry from cold water fish farm centers in Mozandaram Province, Iran. Sebastiao *et al.* (2011) conducted research on the hematological alterations in Nile tilapia (*Oreochromis niloticus*) exhibiting spontaneous infections of columnaris disease caused by *F. columnar*. Recent research in this area has demonstrated the impact of additional diets or herbal extracts on many fish species infected with bacterial infections (Ghodratizadeh *et al.*, 2011; Hoseinifar *et al.*, 2011). Since then, a substantial body of work has been amassed in the literature regarding the various physiological responses of fish, primarily focusing on baseline or comparative values of blood and serum parameters in healthy fish and those subjected to metal or pesticide toxicants (Menon 1952; Bhaskar Rao and Rao 1985; Joshi 1989; Grover *et al.*, 1993; Rauthan *et al.*, 2001 and 2012; Sateesh Kumar *et al.*, 2012). Conversely, research pertaining to alterations in the haematological parameters of fish in response to diverse infectious disorders, particularly microbial infections, is deficient, and the existing literature predominantly focuses on parasitic infections. Research on modifications in hematological and biochemical parameters resulting from parasitic infections mostly focuses on alterations produced by ectoparasitic infections from copepod and monogenean parasites in freshwater fish species. *Channa punctatus* Nair and Nair (1983), *H. Fossilis*, Saxena, and Chauhan (1993), *Clarias batrachus*. Sinha (2000) and Harris *et al.* (2005) discuss the fish ecology of Anchar Lake, Kashmir, namely *Channa punctatus*. Shah *et al.* (2009), Nandus nandus; Kaur *et al* . (2012);and *Cyprinus carpio* (Haidar Ali and Ansari, 2012).

Recent investigations were initiated to examine the impact of feed additives, such as probiotics or herbs, on the hematological and biochemical parameters of various fish species experimentally inoculated with bacterial and fungal pathogens (Das *et al*., 2006; Kumar *et al* ., 006 and 2008; Harikrishnan *et al*., 2003 and 2005; Nayak *et al*., 2007; Jha *et al*., 2007; Harikrishnan *et al*., 2010; Kaleswaran *et al*., 2012). The influence of a herbal extract on the biochemical responses of C. was studied. Dahiya *et al*. (2012) documented the impact of three probiotics on the hematological parameters of Indian magur, *Clarias batrachus*, inoculated with two bacterial pathogens. The aforementioned literature indicates a scarcity of information from the Indian region regarding the physiological responses of fish under diseased conditions, whether arising from natural or experimental infections. Furthermore, research of this kind will facilitate the generation of data that is crucial for mitigating disease issues in culture ponds and for developing innovative diagnostic techniques for assessing fish health. It is indeed astonishing that no studies exist on these features of carps from Indian regions, which are extensively cultivated and considerably contribute to the national economy. Numerous studies exist on the toxicity of metals, bacterial and fungal diseases, pesticides, heavy metals in native fish, and environmental contaminants on the haematological parameters of *Channa striata* and *Channa punctata* from Hasanparthy, Bandham, and Dharmasagar Lakes.

Zarejabad *et al*. (2010) investigated the impact of variations in environmental temperature on the haematological and biochemical parameters of Hoso hoso juveniles. This outcome indicated that haematocrit, calcium, and eosinophil levels were influenced by varying temperatures. Rising temperatures result in a substantial increase in haematocrit, calcium, and eosinophils, whereas there is a minor decrease in white blood cell lymphocytes, cortisol, and glucose concentrations. The haematocrit, hemoglobin concentration, and mean cell hemoglobin concentration data reflect the blood's oxygen-carrying capacity in teleost fish. The primary indices of red blood cells may exhibit significant variability and vary by fish species (Roy George *et al*., 2015). Nevertheless, the information regarding the haematological and biochemical responses of fish to bacterial infections is quite limited in the Indian context, with most available data derived from recent studies. Ghosh and Homechaudhuri (2010) analyzed the blood parameters of *Channa punctatus* after artificial inoculation with A. Salmonicida and A. Hydropila. Haniffa and Mydeen (2011) have conducted studies on total RBC and WBC counts, PCV, Hb% values, and biochemical parameters, including total proteins and enzyme activities in *Channa striatus* experimentally infected with A. Hydrophila. This illustrates that the haematological indices of specific freshwater fish species, such as *Channa striatus* and *Channa punctatus*, were affected by environmental stressors, including the presence of heavy native fish, low oxygen levels, and both organic and inorganic substances (Folmar, 1993).

There are significant variations in the hematocrit and Mean Corpuscular Volume (MCV) among individual healthy fish. Fagbenro *et al.* (2000) investigated the haematological profile, dietary composition, and enzyme activity in the gastrointestinal tract of the African bony-tongue fish, Heterotis niloticus. Hrubec *et al.* (2001) investigated age-related alterations in hematological and plasma biochemical parameters of hybrid striped bass (*Morone chrysops* and *Morone saxatilis*). The results indicated that the packed cell volume and red blood cell indices were much lower, whereas plasma protein concentration was significantly greater in younger fish. White blood cell and lymphocyte levels were dramatically elevated in fish aged six and nine months. Orun *et al.* (2001) conducted a study to assess and compare the blood parameter levels of *Alburnoides bipunctatus, Chalcalburnus mossulensis*, and *Cyprinion macrostomus*. The total leukocyte, neutrophil, and monocyte counts were seen to be elevated in female fish, particularly during the reproductive season, compared to male fish. Haemoglobin, haematocrit, and erythrocyte levels were elevated in male fish over the course of a year. The red blood cell indices were elevated, enabling the species to operate normally even in suboptimal oxygen levels in its natural habitat. The counts of lymphocytes and monocytes were reduced. Svetina *et al.* (2002) investigated the hematology and other blood chemistry parameters of juvenile carp up to three years of age. The findings indicated that the examined haematological parameters might effectively be employed in assessing the metabolic equilibrium and health condition of fish (Sowunmi, 2003).

MATERIALS AND METHODS

Acquisition and Preservation

Control and infected *Channa striatus* and *Channa punctatus* exhibiting EUS were collected without regard to age, sex, or size from Dharmasagar, Hasanparthy, and Bandham lakes, situated 30 km from the Kakatiya University campus, during the rainy and winter seasons from July 2022 to February 2024. The specimens were sourced from local freshwater bodies (tanks, ponds, and lakes) and subsequently transported to the laboratory, where they were housed in large glass aquaria. Only healthy and infective fish were utilized for the experiments. Fish of average size, measuring between 15 and 25 cm in length and weighing from 0.5 grams to 1.0 kilograms, were picked for this study. During the experiment, the laboratory fish were nourished with groundnut cake and sliced goat liver.

Blood Collection

A small volume of blood was extracted from the caudal peduncle using a disposable sterile syringe (2.5 ml) and subsequently transferred to a 0.5 ml miniplast tube containing Di Potassium (EDTA) (1.26 mg/0.6 ml) as an anticoagulant for hematological analysis, as delineated by Hrubc and Smith (2001). RBC and WBC counts were determined using a Haemocytometer, yielding values of 10^6 mm^3 and 10^3 mm^3, respectively, according to Wintrobe (1967). The hemoglobin content was assessed utilizing a hemoglobin test kit (DIAGNOVA, Ranbaxy, India). Blood samples for the total erythrocyte and total leukocyte counts were taken according to the procedure established by Wintrobe (1976). EDTA (Ethylene Diamine Tetraacetic Acid/Disodium salt) was utilized as an anticoagulant. Blood samples were obtained in the morning to mitigate diurnal fluctuations for the estimation of haematological parameters (Gayatri *et al*., 2014). Blood for hematological tests was collected using the procedure of Steuke and Schoettger (1970) by slicing the caudal peduncle.

The obtained blood samples were promptly analyzed hematologically. Clinical evaluations of the fish were conducted in accordance with the method described by Austin *et al*. (2007). The hematological parameters and indices (Red Blood Cell count, White Blood Cell count, Differential Leukocyte Count, Hemoglobin percentage, Packed Cell Volume, Mean Corpuscular Volume, Mean Corpuscular Hemoglobin, and Mean Corpuscular Hemoglobin Concentration) were determined using standard formulas from Campbell (2004). The blood was diluted with suitable diluents for RBC and WBC counts and assessed using an upgraded Neubauer haemocytometer, calculated according to Blaxhall and Daisley (1973). Replicated measurements were conducted for each blood sample to reduce error. A blood film was produced and stained with Giemsa stain for morphological analysis, micrometry, and differential leukocyte count. The haematocrit (Hct) was measured using the Microhaematocrit method (Snieszko, 1960). The Mean Corpuscular Volume (MCV) was computed in accordance with Feldman *et al*. (2000). Mean Corpuscular Hemoglobin (MCH) and Mean Corpuscular Hemoglobin Concentration (MCHC) were computed as per Stoskopf (1993).

Hematological Parameters

The hematological parameters, including Hemoglobin (HB %), Total Erythrocyte Count (TEC), Total Leukocyte Count (TLC), Differential Leukocyte Count (DLC), Packed Cell Volume (PCV), Mean Corpuscular Volume (MCV), Mean Corpuscular Hemoglobin (MCH), and Mean Corpuscular Hemoglobin Concentration (MCHC), were assessed to evaluate the Hematological Index (HI) for understanding the quality of the aquatic environment within the context of fish clinical biology.

Hemoglobin Concentration (Hb)

The haemoglobin concentration in the blood of *Channa striatus* and *Channa punctatus* was determined using Sahli's method, also referred to as the Acid haematin method. In this procedure, haemoglobin is converted to acid haematin through the addition of N/10 or 0.1N HCl, and the resultant brown hue is compared with a standard solution. A Sahli's haemoglobinometer (Germany) was employed for the haemoglobin quantification, with N/10 HCl filled to the 10.0 mark in the graduated tube. A blood sample (20 cu. mm. or 0.02 ml) collected from the fish was promptly transferred into the graduated tube. The tube was gently shaken and allowed to stand for 10 to 20 minutes. Subsequently, N/10 HCl was added dropwise, with mixing after each addition, until the blood color matched the standard hue. The results were recorded from the graduated tube scale, and haemoglobin concentration was expressed as grams per cent.

Total Erythrocyte Count (TEC)

The total erythrocyte count was estimated using a Spencer Bright line counting chamber (Germany) with an enhanced Neubauer rule. Blood was taken in a sample vial containing 2% EDTA as an anticoagulant. Blood was extracted from the sample vial into the RBC pipette precisely to the 0.5 mark. The pipette containing blood was immersed in a red blood cell diluting fluid and filled to the 101 mark (Hayem's diluting fluid from BDH firm was utilized in this study). Consequently, the blood dilution was 1:200. The blood and fluid in the pipette were meticulously blended by gently spinning the pipette between the fingers. A clean, grease-free Neubauer slide was positioned on the microscope's mechanical stage, followed by the placement of a clean coverslip over the slide. The counting chambers are shown under adequate illumination using low power. The initial two drops of blood in the pipette were eliminated, as they do not contribute to the dilution. By gently compressing the rubber tube of the pipette, blood was expelled, and a drop was positioned beneath the coverslip while the pipette was held at a 5° angle. The drop of blood disseminated gradually and uniformly under the coverslip due to capillary action. It was permitted to remain undisturbed for 2 to 3 minutes for the erythrocytes to settle entirely. The red cells manifested as distinctly defined particles, while the chamber lines seemed black. The erythrocytes were enumerated in five squares, each including 16 smaller squares, with four located at the corners and one at the center. The total erythrocyte count per cubic millimeter of blood was derived from the subsequent computation.

The Total Count of Large Squares is 5

Total quantity of smaller squares (each larger square encompasses)

Sixteen little squares equals 16 multiplied by 5, resulting in 80.

The aggregate count of RBC over all five squares equals X.

The area of the smallest square is 1/400 square units. Hmm.

The depth of the counting chamber is 0.1 mm.

The volume of the smallest square is calculated as 1/400 multiplied by 1/10, resulting in 1/4000 cubic millimeters.

Blood dilution = 1:200 (1/200).

Count of cells multiplied by dilution factor multiplied by depth factor

Total quantity of snail squares

(X x 200 x 4000)/80 = millions of RBCs per cubic millimeter of pure blood.

Total Leukocyte Count (TLC)

The approach for estimating the total leucocyte count was identical to that used for the total erythrocyte count. The amount of white blood cells was significantly lower than that of erythrocytes. Consequently, the necessary dilution was significantly reduced. A specialized pipette, smaller than the RBC pipette, was utilized. It was calibrated at 0.5, 1, and 11, allowing for a blood dilution ratio of 1:20. Turk's fluid served as the diluent.

Blood was drawn to the 0.5 mark in the WBC pipette and diluted to the 11 mark with diluent. The blood and diluent were thoroughly mixed, and after discarding the initial two drops, a drop was placed under the coverslip of a Neubauer slide and permitted to spread and settle. The white blood cells manifested as black spots surrounded by a bright white region. Leucocytes were enumerated in each of the four corner squares. The overall cell count was determined in thousands per cubic millimeter.

The total count of corner squares is 4.

Total quantity of WBCs in four squares = X

The mean quantity of white blood cells in each corner square is equal to X divided by 4.

Length of one square equals 1 mm.

The width of one square is 1 mm.

Area of a single square = 1 square millimeter.

Depth = 0.1 mm.

The volume of each corner square is 1/10 cubic unit. Hmm.

Dilution at a ratio of 1:20

Count of cells multiplied by dilution factor multiplied by depth factor

Count of squares enumerated.

(X x 20 x 10)/804 = thousands of WBCs per cubic millimeter of pure blood.

Differential Leukocyte Count (DLC)

The differential leukocyte count is a technique used to quantify the various types of white blood cells present in a blood sample by analyzing a well-stained blood smear. The quantity of each white blood cell type is represented as a percentage of the total cell count, typically 100. For the differential leukocyte count, a blood sample was obtained on a grease-free slide from the individual fish, and a thin smear was formed within 10 to 20 seconds following the incision. The film was promptly air-dried, then fixed in methyl alcohol, and subsequently stained with Wright's, Irishman's, and Giemsa's stains. Among all of them, Giemsa's stain yielded favorable findings. The counts were conducted using oil immersion magnification as detailed by Khan and Siddiqui (1969) and Agrawal and Srivastava (1980). The proportion of leucocytes in an individual fish was determined by counting 100 cells from the best of ten smears, and the average of these ten counts was calculated. Larger leucocytes tended to be located near the periphery of the smear, whilst smaller leucocytes were predominantly found in the center. Consequently, the counts were conducted to ensure that all edges and the middle area were encompassed. The identified cells were classified as lymphocytes, monocytes, eosinophils, neutrophils, and basophils.

Packed Cell Volume (PCV)

The PCV is defined as the volume of packed red blood cells in a blood sample, represented as a percentage of the total blood volume. The packing of red cells is accomplished by centrifuging a blood column in a glass tube. Uncoagulated blood was subjected to centrifugation in Wintrobe's graduated pipette (Haematocrit tube) at 3000 rpm for 30 minutes. The corpuscles were consolidated to a uniform volume after 30 minutes. The volume of packed cells was thereafter expressed as a percentage of the total blood volume in the hematocrit tube. This is sometimes referred to as the hematocrit value.

Hematological Indices

Following the assessment of erythrocyte count, hemoglobin concentration, and packed cell volume, many formulas were derived, providing quantitative data regarding red blood cells. The values referred to as blood indices were computed in the current experiments as follows:

Mean Corpuscular Volume (MCV)

It is the mean volume of an individual red blood cell, given in cubic microns.

MCV = (Packed Cell Volume x 10) / (RBC in millions per cubic millimeter)

Mean corpuscular hemoglobin (MCH)

Mean corpuscular hemoglobin denotes the average weight of hemoglobin, measured in micrograms, contained within a single cell. It is affected by the cell's size and the hemoglobin concentration.

MCH = (Hemoglobin concentration x 10) / (RBC in millions per cubic millimeter)

Mean Corpuscular Hemoglobin Concentration (MCHC)

Mean corpuscular hemoglobin concentration denotes the average concentration of hemoglobin within red blood cells. MCHC is unaffected by cell size, unlike MCH. This is articulated as a percentage.

MCHC = (Hemoglobin concentration x 100) / (Packed cell volume)

OUTCOMES

Statistical Examination

The results were derived with SPSS (12.0) for Windows. Each assay was conducted in quintuplicate. Values were presented as Mean ± SD of replication at $p<0.0001$ for infected (T) and control (C) fish. The literature concerning fish haematology is frequently perplexing due to the inconsistency in morphological criteria and terminology used for blood and cell identification. This contributed to the challenges encountered in the current study. Observations of the haematological values obtained from *Channa striatus* and *Channa punctatus* in both healthy and EUS-infected fish were conducted. The haematological study has emerged as a crucial instrument for fishery research. Blood values significantly varied among several freshwater fish species, which is believed to indicate adaptations to diverse environmental conditions (Ramaswamy *et al.*, 1978; Moyle *et al.*, 1982). The findings indicate that the levels of hemoglobin, packed cell volume, mean corpuscular hemoglobin, mean corpuscular hemoglobin concentration, and the quantity of red blood cells were markedly reduced in *Channa striatus* fish injected with A. The present study presents the haematological parameters, including RBC, WBC, HB, PCV, MCV, MCH, MCHC, and Differential Leukocyte Count, in comparison to control fish, as reported by Haniffa *et al.* (2011) (Tables **1** and **2**). Haematological parameters of control and infected Channa striata are provided in Table **1**.

Table 1. Haematological parameters of control and infected *Channa striatus*.

S. No.	Blood Parameter	Control Fish	Infected Fish	Percentage Change (%)	P Value
1	HB (g/dl)	11.3±0.34	8.7±0.31	-91.3	0.0001
2	PCV %	26.3±0.35	11.95±0.20	-88.5	0.0001
3	MCV (M.gms)	71.7±1.22	52.6±0.70	-47.4	0.0001
4	MCH (M.gms)	23.9±0.40	22.0±0.29	-78	0.0015
5	MCHC(M.gms)	33.1±0.56	35.6±0.47	-64.4	0.0033
6	RBCs ($10^{6/ml}$)	4.4±0.25	1.96±0.14	-98.04	0.0001
7	WBCs ($10^{3/ml}$)	41670±947.51	50375±893.85	50275	0.0001
8	Lymphocytes %	52.0±0.71	57.4±0.80	-42.6	0.0001
9	Neutrophils %	59.6±0.62	28.9±0.65	-71.1	0.0001
10	Monocytes %	5.7±0.34	3.4±0.31	-96.6	0.0001
11	Eosinophils %	2.5±0.37	4.0±0.36	-96	0.0106
12	Basophils %	2.3±0.26	1.7±0.35	-98.3	NS

Results were statistically analyzed by Student's t-test, and values presented in the table are Mean± SD.

Haematological parameters of control and infected *Channa punctatus* are presented in Table **2**.

Table 2. Haematological parameters of control and infected *Channa punctatus*.

S. No.	Blood Parameters	Control Fish	Infected Fish	Percentage Change (%)	P Values
1	HB (g/dl)	10.1±0.33	9.1±0.29	-90.9	0.0373
2	Haematocrit (PCV) %	32.7±0.68	28.3±0.47	-71.7	0.0001
3	MCV (M.gms)	56.7±0.94	65.7±1.32	-34.3	0.0001
4	MCH (M.gms)	21.4±0.61	18.9±0.31	-81.1	0.0015
5	MCHC (M.gms)	36.539±0.925	33.159±0.174	-66.841	0.0021
6	RBCs ($10^{6/ml}$)	4.2±0.27	1.89±0.13	-98.11	0.0001
7	WBCs ($10^{3/ml}$)	33505±674.13	39000±756.450	38900	0.0001
8	Lymphocytes %	58.2±0.341	61±0.614	-39	0.0009
9	Neutrophils %	34.8±0.64	29.3±1.67	-70.7	0.0059
10	Monocytes %	5.8±0.37	4.4±0.74	-95.6	0.0057
11	Eosinophils %	2.6±0.37	4.1±0.38	-95.9	0.0124
12	Basophils %	2.2±0.28	2.8±0.38	-97.2	NS

Results were statistically analyzed by Student's t-test, and values presented in the table are Mean± SD.

Erythrocytes (RBC)

The data indicate a substantial reduction in RBC count. The Total Erythrocyte count was noted to be diminished in EUS-infected *Channa striatus* and *Channa punctatus* by 63.5% and 55%, respectively, compared to the control. A reduction in RBC was also documented by Malathi *et al* . 2012). The sedentary behavior of these fish, coupled with fungal infection, leads to blood extravasation and a decrease in hemogenesis, ultimately impairing the hematopoietic tissue's ability to release blood cells (Xavier *et al.*, 2011). Erythrocytes, or red blood corpuscles, are the primary cellular components present in the blood. The observations indicated that the morphology of a normal red blood cell, or normocyte, varies in shape, presenting as oval, elliptical, biconcave disc, or round.

The nuclei are spherical and located at the center of the cells. No cellular inclusions or vacuoles are present in the cytoplasm. Examination of living red blood cells in the counting chamber of a hemocytometer under a microscope revealed a light brown coloration of the cells, with a distinctly visible nucleus. In a stained blood film, the nucleus and cytoplasm of the red blood cells displayed variable colors with different stains. The blood smear stained with Leishman's stain revealed a violet nucleus and a blue cytoplasm. Conversely, staining with

Wright's stain produced a markedly different coloration of the red blood cells. The nucleus had a pink hue against a reddish-yellow cytoplasmic backdrop. In the Giemsa-stained blood film, the nucleus appeared violet to blue, with only the cytoplasmic borders displaying blue, resulting in a white center region. The observed average size of the red blood cells is 10.75 × 10.35 microns. Fish infected with bacteria and fungi exhibited no significant alterations in the morphology or cytological inclusions of the erythrocytes. The total erythrocyte count recorded in control *Channa striatus* and *Channa punctatus*was 4.4 ± 0.25 x 10^6/cu and 4.2 ± 0.27 x 10^6/cu, respectively. In diluted blood, fish infected with bacteria and fungi had a reduction in erythrocyte counts, recorded at 1.96±0.14 x 10^6/cu. 1.89 ± 0.13 x 10^6/cu. Hmm. Signifying a reduction of around 63.5% and 55% in red blood cells. A notable decrease in erythrocyte count is observed in EUS infection; nonetheless, it remains at 36.5%, which is 45% lower than the control value. Blood smear of control fish (*Channa striatus*) is shown in Fig. (**1**).

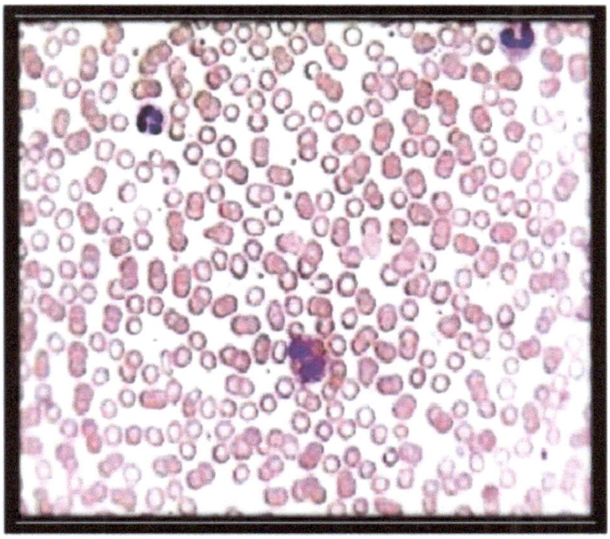

Fig. (1). Blood smear of control fish (*Channa striatus*).

Hemoglobin (Hb)

Hemoglobin is a complex compound found in red blood cells that imparts a red hue to blood. It is an iron-containing protein and serves as a respiratory pigment. The primary purpose of hemoglobin is to transport oxygen to tissues and facilitate the removal of carbon dioxide from them. The oxygen-carrying ability of blood mostly depends on its hemoglobin content. Hemoglobin is transformed into acid hematin in red blood cells to assess the hemoglobin level.

The hemoglobin content of *Channa striatus* and *Channa punctatus* was 11.3±0.34 g and 10.1±0.33 g, respectively, in control fish. In contrast, fish infected with bacteria and fungi exhibited hemoglobin levels of 8.7±0.31 g and 9.1±0.29 g. Compared to the control, the hemoglobin percentage in infected fish decreased by 23% and 9%, respectively. The variation in hemoglobin (Hb) levels in fish from infected lakes is attributed to seasonal pollution, low oxygen levels, the presence of large native fish, and chemical stress. Mahajan *et al.* (1979) and Devi *et al.* (2004) reported a reduction in Hb content in Heteroneustes fossils exposed to paper mill effluents. Fungal and bacterial infections adversely affect the hematopoietic system (Xavier *et al.*, 2011). In accordance with the reduction of RBC count resulting from bacterial and fungal infections, the hemoglobin percentage in fish also showed a comparable decline.

Packed Cell Volume (PCV)

A significant reduction in PCV in *C. Striatus* and *C. Punctatus* exhibited PCV values of 55% and 13.5% in EUS-infected fish. The reduction in PCV values unequivocally signifies the presence of anaemia in the infected fish from Dharmasagar, Hasanparthy, and Bhandham lakes. Srivastava *et al.* (1981) found variations in packed cell volume while examining seasonal and diurnal fluctuations in blood cell types of the freshwater teleost, Colisafasciatus.

Mean Corpuscular Volume (MCV)

Mean Corpuscular Volume in infected fish was seen to decrease by 27% and 15% compared to the control group *C. Striatus* and *C. Punctatus* imply a condition of microcytic anemia resulting from EUS infection. A reduction in MCV in freshwater fish is attributed to aberrant stress conditions, including seasonal, environmental, and toxic factors (Weils *et al.*, 1983; Benarjee *et al.*, 1986). Chronic fungal, bacterial, and viral infections result in a dramatic reduction of MCV levels in fish species such as *Saccobranchus fossilis*, leading to a disease known as normochromic microcytic anemia (Verma *et al* ., 1979).

Mean Corpuscular Hemoglobin (MCH)

In comparison to the control, the MCH values in the EUS-infected *Channa striatus* and *Channa punctatus* fell by 8% and 15%, respectively, indicating hypochromic anemia. The current results align with the research conducted by Benarjee *et al.* (1986), which examined the toxicological effects of organophosphorus insecticides on histological alterations in *Clarias batrachus*.

Mean Corpuscular Hemoglobin Concentration (MCHC)

In comparison to the control, EUS-infected *Channa striatus* and *Channa punctatus* have demonstrated a rise in MCHC levels of 7.5% and 9.2%. The elevation in MCHC value is attributable to macrocytic anemia.

Total Leukocyte Count (TLC)

A notable increase of 21% and 17% in total leucocyte count was observed in the infected fish. The significant elevation in leucocyte count in infected *Channa striatus* and *Channa punctatus* can be attributed to leukocytosis resulting from anoxic stress, low temperatures, food scarcity, elevated organic and inorganic chemicals (NH_3), and the presence of heavy native fish (Gautham *et al.*, 1989). Research on *Channa striatus* and *Channa punctatus* has indicated that leucocytes represent the second most significant category of circulating blood cells and demonstrate enhanced proliferation upon exposure to mitogens, such as concanavalin, as well as infections from Aphnomyces invadans, Aspergillus fumigatus, and *Aspergillus niger*, suggesting that leucocytes are polyclonally activated.

DIFFERENTIAL LEUKOCYTE COUNT (DLC)

Lymphocytes

The current investigation indicates that the lymphocyte percentage in the control groups *Channa striatus* and *Channa punctatus* was 52.0±0.71% and 58.2±0.341%, respectively. A notable increase of 5.4% and 5% was seen in the diseased fish population. This syndrome is termed lymphocytosis, an immune reaction to stress generated by environmental pollution. An elevation of lymphocytes signifies a pathogenic condition resulting from EUS and toxic stressors in the infected fish (Laxmareddy *et al.*, 2013). An increased lymphocyte proliferation was also noted in Atlantic Manhadevan with ulcer disease syndrome (Faisal *et al.*, 1992). Leukocytes of *Channa striatus* are also shown in Figs (**2** - **4**).

Neutrophils

In the current study, neutrophils comprise a significant proportion of the leukocyte count in the control fish. The neutrophil percentage in the control fish of *Channa striatus* and *Channa punctatus* reported values were 59.6 ± 0.62% and 34.8 ± 0.64%. In the EU, the prevalence of contaminated fish reduced dramatically by 30.7% and 18%. This disorder is frequently referred to as neutropenia. All physiological causes of leukopenia result in neutropenia. Hematological values vary in response to environmental changes, as shown by

Mohan *et al.* (1989). Differential Leukocyte Count (DLC) in control and infected *C. striatus* and *C. punctatus* fish is shown in Figs. (**5** and **6**).

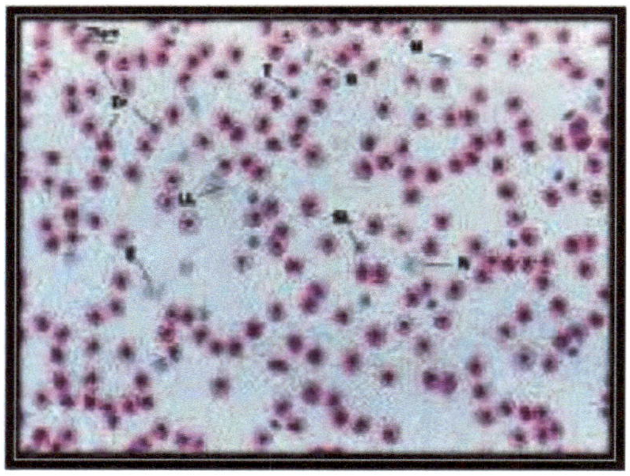

Fig. (2). Leukocytes of *Channa striatus*.

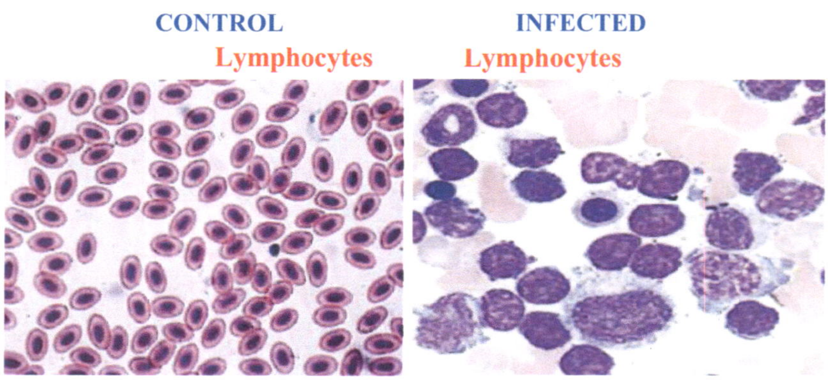

Figs. (3 and 4). Differential Leukocyte Count (DLC) in control and infected *Channa striatus* and *Channa punctatus* fish.

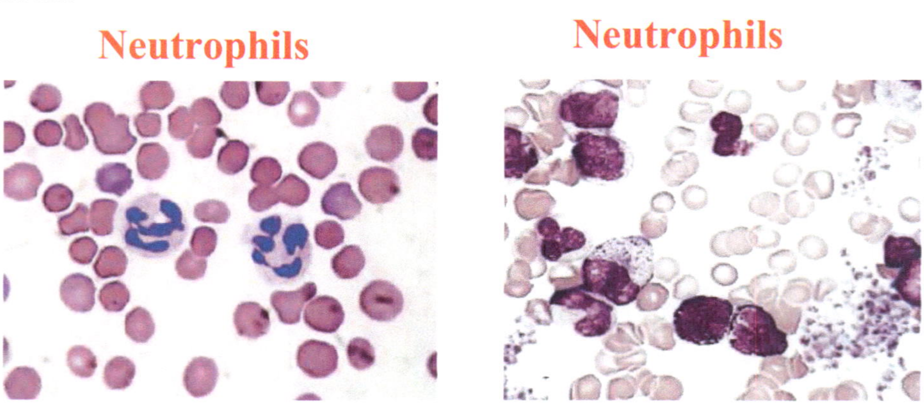

Figs. (5 and 6). Differential Neutrophil Count in control and infected *C. striatus* and *C. punctatus* fishes.

Monocytes

The monocyte percentages were 5.7±0.34% and 5.8±0.37%. Within the control, *C. Striatus* and *C. Punctatus* fishes exhibited a reduction to 2.3% and 2.4% in the presence of bacterial and fungal infections, respectively. Similarly, these data indicated large increases in monocytes, as reported by Thakur *et al*. (1990), in the infected air-breathing fish, *Clarias batrachus* (Plate IX, Figs. (**7** and **8**).

Monocytes Monocytes

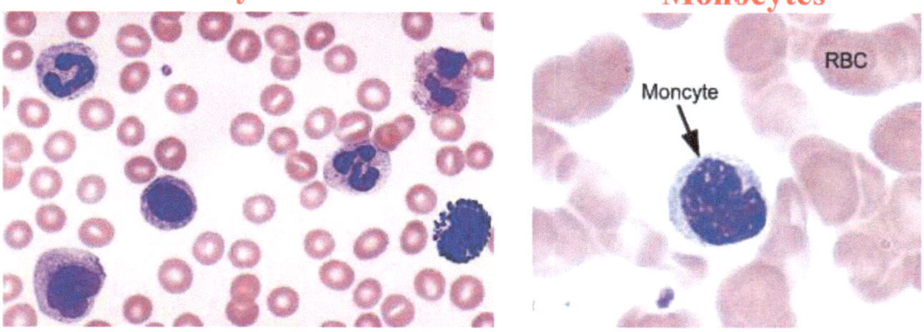

Figs. (7 and 8). Differential monocyte count in control and infected *C. striatus* and *C. punctatus* fishes.

Eosinophils

Eosinophils are fewer in number than neutrophils. The results indicate that the average eosinophil count was 2.5±0.37% and 2.6±0.37. In the control group of *C. striatus* and *C. punctatus*, there was a substantial rise of 1.5% and 5.8% in the infected fish, respectively. The elevation in eosinophil percentage may be attributable to EUS infection. Water pollution adversely affects blood parameters, thereby elevating eosinophil counts due to the induction of allergies; analogous findings were noted in infected *Clarias batrachus*, an air-breathing fish (Thakur *et al*., 1990). *C. striatus* and *C. punctatus* fish control and Infected Differential Leukocyte Count (DLC) are shown in Figs. (**9** and **10**).

Eosinophil's Eosinophil's

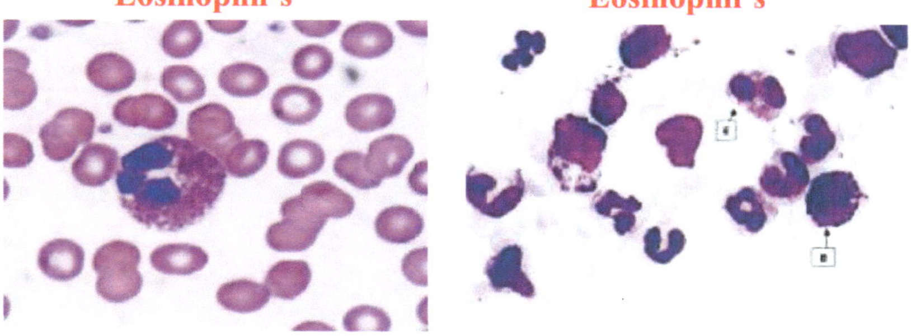

Figs. (9 and 10). *C. striatus* and *C. punctatus* fish control and Infected Differential Eosinophil's count.

Basophils

The standard documented basophil percentages for the control *C. Striatus* and *C. Punctatus* fishes exhibited values of 2.3±0.26% and 2.2±0.28%. In fish infected with bacteria and fungi, the levels have decreased to 0.6% and 0.8%, respectively. The current results align with the findings of Thakur *et al*. (1990) for infected *Clarias batrachus*. *C. striatus* and *C. punctatus* fish control and Infected Differential Leukocyte Count (DLC) are shown in Figs. (**11** and **12**).

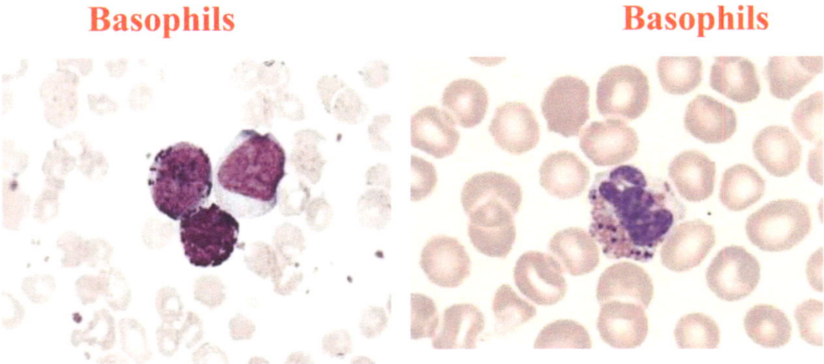

Figs. (11 and 12). *striatus* and *C. punctatus* fish control and infected differential basophils count.

DISCUSSION

Hematological tests have long served as crucial diagnostic tools in medicine, and recent studies suggest they may also be significant indicators of disease in freshwater ecosystems. Hematological alterations in fish are associated with temperature, seasonal fluctuations, inadequate nutrition, and exposure to bacterial, fungal, pesticide, protozoan, and trematode parasites, as well as environmental stressors. Similar to other vertebrates, fish blood is red and comprises a suspension of blood cells within the plasma. The plasma consists of a solution of proteins and electrolytes, whereas the blood cells include erythrocytes and leukocytes. Blood is the most vital bodily fluid, and its composition typically indicates the physiological state of the body. Cassilas and Smith (1977) assert that the hematological parameters of fish fluctuate with environmental conditions and promptly react to alterations in water quality. Haematological data are increasingly utilized to examine pathological conditions. The total erythrocyte count in *Channa striatus* and *Channa punctatus* was significantly diminished, resulting in anaemia, when exposed to various bacteria and fungi, including *Aeromonas hydrophila, Staphylococcus aureus, Pseudomonas aeruginosa, Salmonella salmonicida*, and *Ascomycota species* such as *Aspergillus flavus, Fusarium solani, Rhizophus stolenifer, Aspergillus fumigatus, Penicillium chrysogenum, Aspergillus niger* and *Trichoderma viride. Escherichia coli* and *A.*

Hydrophila exhibited the most significant reduction in fish infected with EUS, followed by *Pseudomonas aeruginosa*. The greater reduction in the initial two instances may result from the synergistic interaction of various bacterial species. The pathogenicity of bacterial and fungal species may result from the synthesis of haemolysin, which lyses and obliterates erythrocytes. Of the two fish species examined, *Channa striatus* exhibited greater susceptibility to bacterial infection than *Channa punctatus*. Additionally, seasonal variations introduced diverse bacterial species, which collectively impacted both fish, resulting in a significant decrease in red blood cell count.

The erythrocyte count in *Channa striatus* exhibits a strong correlation with haemoglobin and haematocrit concentrations, while showing an inverse relationship with leucocyte count. In contrast, *Channa punctatus* demonstrates a moderate correlation with haemoglobin, a strong correlation with haematocrit, and a comparatively weaker correlation with leucocytes. Anaemia linked to A. has been documented. Hydrophilic infection may generate protease enzymes that could lyse erythrocytes, leading to a decrease in RBC count in *C. Carpio* (Harikrishnan *et al.*, 2003) and in goldfish, *Carassius auratus* (Brenden and Huizinga, 1986b). Wedemeyer *et al.* (1984) suggest that the decrease in erythrocyte count may be attributed to haemodilution caused by compromised osmoregulation across the gill epithelium. The reduction in RBC may also result from the detrimental impact on the haemopoietic process, causing microcytic, poikilocytotic, and haemolytic anaemia. Erythropoietin facilitates erythropoiesis. Bhatt and Farswan (1992) and Bala and Bais (2000) indicated that the reduction of erythrocytes in several fish species may result from the hypostimulation of erythropoietic tissue under stress conditions. The current study observed an early decline in RBC count throughout the winter season in *C. striatus* and in *C. Punctatus* for all bacterial infections, which is succeeded by a notable and rapid decrease.

This pattern may result from increased erythrocyte production to mitigate initial stress, followed by subsequent degradation. Barham *et al.* (1980) reported that bacterial infections caused by Aeromonas and Streptococcus led to increased fragility of erythrocytes and elevated ESR readings. Brenden and Huizinga (1986a) asserted that the endotoxin of Aeromonas species had significant pathogenic effects in mice, resulting in hypochromic microcytic anemia. Waagbo *et al.* (1988) noted a comparable decrease in RBC in Atlantic salmon. Prasad and Qureshi (1998) documented erythrocyte lysis and reduced hemoglobin levels in Cirrhinus mrigala infected with bacterial gill infections. Comparable results were observed in fish examined for bacterial infections (Bruno 1986; Prasad 1992; Suzumoto *et al.*, 1997). Fish afflicted by Epizootic Ulcerative Syndrome (EUS) exhibited a significant reduction in erythrocyte count. Blaxhall (1972), Yildiz

(1998), Mastan and Qureshi (2002), and Celik and Aydin (2006) have noted that anaemia may be attributable to membrane damage of Red Blood Cells (RBC).

In the current study, the White Blood Cell (WBC) counts of EUS-infected fish increased, peaking in the winter season for C. striatus and C. punctatus, after which the count fell, particularly in relation to bacterial and fungal associations. This exemplified the fish's adaptive response to mitigate stress until the conclusion of the winter season and the ensuing decline. C. Punctatus exhibits greater resistance to bacterial and fungal infections than C. Striatus, demonstrating the varied immunological sensitivity of leukocytes. An elevation in leucocytes may result from stimulated lymphopoiesis or augmented lymphocyte release from lymphomyeloid tissues. This lymphocyte response could be attributed to hypoxic conditions, low temperatures, chemical toxins, or may be linked to bacterial toxins, as documented by Haniffa (1990). This study demonstrates the degree of resistance exhibited by the fish against germs and fungi.

The potential influence of leukocyte activation and proliferation on the formation of granulomatous responses related to the disease is still to be clarified. Catla infected with EUS exhibited anemia, as indicated by a reduced total red blood cell count, packed cell volume, and hemoglobin concentration. Anaemia may result from hemodilution due to fluid loss from hemorrhagic or necrotic lesions in infected catla, as noted in EUS-affected Asian cichlid fish *Etroplus suratensis* (Pathiratne and Rajapakshe, 1998). The elevated total leukocyte counts in EUS-infected catla could account for the heightened immunological responses; the leukocyte count was greater and the hemoglobin content lower in blood from EUS-affected Indian fish compared to healthy fish (Kar *et al.*, 2000). Increased concentrations of antibodies to A. Invasions have been documented in striped snakehead (Channa striata) and rainbow trout (*Oncorhynchus mykiss*) (Thompson *et al.*, 1997; Thompson *et al* 1999; Miles *et al* 2001).

According to Christensen *et al.* (1978), the haemoglobin content of fish serves as a valuable indicator of fish health. Haemoglobin is essential for the survival of fish due to its direct involvement in the oxygen-binding capacity of the blood. The current investigation demonstrated a notable decrease in haemoglobin levels and the oxygen-binding capacity of the blood for *Channa striatus* and *Channa punctatus* in all infectious diseases. Haemoglobin levels diminished more rapidly in bacterial and fungal infections. This may result from the cumulative stress effects of many bacterial and fungal species. *Channa striatus* and *Channa punctatus* persisted through all bacterial and fungal diseases over the seasonal period. The concentration of haemoglobin is directly proportional to its oxygen binding capability (Prosser, 1973). The substantial decrease in haemoglobin content may influence the oxygen binding capacity, as indicated by multiple authors (Bonga 1997; Ruane *et al.*, 1999).

Scott and Rodgers (1981) noticed a substantial drop in RBC count and hemoglobin levels in Ictalurus punctatus during A. hydrophila infection. The hydropneumonic infection and the reduction in hemoglobin levels may result from the swelling and rupture of red blood cells, together with inadequate mobilization of hemoglobin from the spleen and other hematopoietic organs. The aforementioned findings corroborated the current view that stress may result in a substantial reduction in hemoglobin levels, potentially causing anemia, which could impair oxygen binding capacity.

In addition to oxygen delivery, erythrocytes perform many other functions in the body. The reduction in erythrocyte count and hematocrit value in the blood may be attributed to the fish's incapacity to manage metabolic stress, resulting in the failure to elevate RBC count to satisfy heightened oxygen requirements (Fukuda, 1997). A notable decrease in erythrocyte count, hemoglobin levels, and hematocrit concentration, together with an increase in leukocyte count, was observed in all bacterial exposures. Haemopoiesis may be significantly impacted during bacterial infections (Barham *et al* ., 980). The pattern of change resembled that observed in the bacterial infection of A. Hydropilin goldfish, *Carassius auratus* L. Brenden and Huizinga (1986b) examined Pseudomonas fluorescens infection in carp, *Cyprinus carpio* L. Yildiz (1998), Carusu *et al.* (2002), and Yildiz and Bekcan (2005) documented a decrease in haematocrit levels in Silurus glanis L. infected with Edwarsiella tarda and in eel, Anguilla anguilla infected with A. Hydrophila and an elevation in the leukocyte count corroborated the current findings. Variations in haemoglobin concentration and erythrocyte count may arise from species differences and seasonal fluctuations (Molnar and Tamassy, 1970).

This study's observations suggested that the bacterial infection with *E. coli* changed most haematological parameters, and A. Hydrophila or alternative combinations of bacterial species. Bacterial and fungal toxins, including endotoxins and exotoxins, caused infections in numerous internal organs such as the skin, gills, liver, pancreas, and kidneys, consequently disrupting hematopoiesis and reducing erythrocyte volume in peripheral circulation.

Besis (1972) posits that pathogenic alterations result in anemic situations, potentially stemming from aberrant erythropoiesis, insufficient hemoglobin functionality, and injury to red blood cells or heightened erythropoiesis by the bone marrow. The elevation of leucocytes, particularly granulocytes, in afflicted fish may result from heightened tissue damage caused by pathogens and various stressors (Iwama *et al.*, 1986), while the reduction in lymphocytes could also be attributed to bacterial infections and other stressful circumstances (Mastan and Qureshi, 2002). The leukocyte population in fish elevates as a defensive response to stress, maybe resulting from increased lymphopoiesis or augmented lymphocyte release from lymphoid tissue (Srivastava 1969; Pickering 1986). Moreover, alterations in the form and size of erythrocytes and leucocytes are the most prevalent morphological abnormalities observed in clinical circumstances.

There was significant heterogeneity in the MCV, MCH, and MCHC among fish, C. Striatus and C. Punctatus, exposed to bacterial and fungal stress. The substantial reduction in these indices in both fish species was inversely related to haematocrit levels and hemoglobin content, resulting in a macrocytic hypochromic anaemic condition. The variation was more significant in C. Striatus and C. Punctatus under bacterial and fungal conditions, whereas E. Fishes infected with E. coli exhibited minimal impact. A. Hydropila infection exhibited effects comparable to those of bacterial consortia, as reported by other authors (Ischikawa 1998; Celik and Aydin, 2006). A significant variation in haematological indices was observed in Black Goby (Selma and Hatice, 2004).

Haney *et al.* (1992) reported that the erythrocytes of infected fish exhibited elevated MCV, reduced MCHC, and diminished corpuscular hemoglobin, corroborating the current findings, which may be attributed to blood loss from lesions caused by bacterial infection. A reduction in MCH was also noted in fish afflicted with parasite illnesses (Kawatsu, 1978). Fluctuations in haematological parameters may account for alterations in indices such as MCH, MCV, and MCHC, which are indicative of acute infection leading to anemia (Svoboda *et al.*, 2001; Jose *et al.*, 2006). The haematological studies performed on experimental fish infected with various bacteria have revealed stress-induced alterations in blood components. The study also demonstrated significant fluctuations in blood parameters during the exposure period in both fish species.

The initial phases of infected EUS fish exhibit no substantial impact on hematological parameters. Nonetheless, chronic stages lead to modifications in the overall hematological profile (Pathiratne and Rajapakshe, 1998). Multiple authors have indicated that the principal erythrocytic indicators, specifically total erythrocyte count, hemoglobin, and hematocrit concentration, are markedly diminished in seriously affected fish, resulting in anemia. Initial hematological investigations indicated reduced hemoglobin levels in the blood of fish affected by EUS compared to their healthy counterparts of the same species (Kar *et al.*, 1994).

The current study illustrates that in the fish species *Channa striatus* and *Channa punctatus*, bacterial and fungal infections significantly alter haematological parameters. Hematological tests of fish are crucial as they are associated with the health of the fish. Fish species are affected by various variables, including pollution, water quality, and microorganisms in their natural habitat. They partially acclimatize to these unfavorable conditions by modifying their physiological functions. Hematological markers were markedly elevated during the infection of A. Hydrophila in infected C. Punctatus (Arulmozhi *et al*., 2010). Reference values for fish species will aid in establishing and identifying the etiologies of fish diseases, a challenge for the ichthyologist. The red blood cell count in fish dictates the capability for transporting dissolved oxygen. The RBC count in healthy *Channa striatus* and *Channa punctatus* exceeds that of fish infected with eus. The sedentary behavior of these fish, coupled with fungal infection, leads to blood extravasation and a decrease in hemopoiesis, ultimately impairing the hematopoietic tissue's ability to release blood cells (Xavier *et al*., 2011). The present investigation demonstrated a decreased red blood cell count in Channa species. This disorder may result from alterations in their habitat or seasonal fluctuations.

Selma and Hatice (2004) identified morphological alterations in the Black Goby, Gobius niger, when subjected to water contaminated by home trash. Kumari and Thomas (2002) observed morphological alterations in leucocytes of Cirrhinus mrigala infected with A. hydrophila. Neutrophils exhibited an increase in size, with the nucleus displaying variability in dimensions and morphology. The monocytes exhibited significant vacuolation (Thomas, 2008). The current investigation showed analogous findings, revealing that stress generated by various bacteria and fungi was varied, resulting in the disintegration of erythrocyte membranes, vacuolation of blood cells, and structural and morphological alterations.

The hemoglobin (Hb) levels in the control groups of *Channa striatus* and *Channa punctatus* were higher than those in the infected fish. Reduced hemoglobin content was observed in those infected with EUS, *C. striatus,* and *C. punctatus.* The fluctuation of hemoglobin resulting from seasonal pollution, diminished oxygen levels, chemical stress from indigenous fish, and fungal and bacterial infections affects the dysfunction of the hematopoietic system (Xavier *et al* , 2011). The diminished hemoglobin and hematocrit levels in EUS-affected Ophicephalus striatus have also been ascribed to hemodilution (Cruz-Lacierda and Shariff, 1995). The low Hct level should be acknowledged, as the discrepancies in blood parameters across fish of different sizes were documented by Satheeshkumar *et al*. (2012) and Fazio *et al*. (2013). The white blood cell counts were significantly elevated in EUS-infected fish compared to the control fish. In all the infected *Channa striatus* and *Channa punctatus* in fish, the differential count of lymphocytes, neutrophils, eosinophils, and monocytes revealed a significant difference between the control and infected specimens. With the exception of the basophils in both fish species, no significant difference was observed ($p < 0.001$).

The substantial elevation of leukocyte count in infected *Channa striatus* and *Channa punctatus* can be associated with leukocytosis resulting from anoxic stress, low temperatures, elevated organic and inorganic chemicals (NH_3), and the presence of heavy native fish species. An elevation in the percentage change of granulocytes (neutrophils and eosinophils) signifies infection in fish (Sahan *et al.*, 2007). The outcome of the PCV in this investigation for *Channa striatus* and *Channa punctata*. The low percentage of Packed Cell Volume (PCV) in EUS-infected fish is likely attributable to the comparatively reduced Red Blood Cell (RBC) counts observed in these specimens. The decline in PCV values unequivocally signifies an anemic state in the infected freshwater fish. Similarly, diminished red blood corpuscles and PCV were documented in coho salmon (Oncorhynchus kisutch) afflicted with Vibrio anguillarum, as noted by Harbell *et al.* (1979) in Asian cichlid fish (*Etroplus suratensis*) suffering from epizootic ulcerative syndrome. Pathiratne *et al.* (1998) studied ulcerous dermatitis in rainbow trout (*Oncorhynchus mykiss*). Rehulka (1998) conducted experiments on rainbow trout infected with Aeromonas sobria, and A. Rehulka (2002) experimentally infected carp (*Cyprinus carpio*) with caviae, *A Hydrophila*. Harikrishnan *et al.* (2003) and in Nile tilapia experimentally infected with Streptococcus iniae. The MCV value indicates the size of red blood cells by quantifying the volume occupied by an individual red blood cell (Chen *et al.*, 2004).

The current investigation demonstrates a markedly reduced MCV value in infected *Channa striatus* and in *Channa punctatus*, which is markedly elevated in sick fish. Infected fish are compared to control fish, which may exhibit decreased Packed Cell Volume (PCV), whereas chronic fungal, bacterial, and viral infections significantly reduce mean corpuscular volume (MCV) values in species such as *Saccobranchus fossilis*, resulting in a condition known as normochromic microcytic anemia (Verma *et al.*, 1979). An elevated MCV is associated with the enlargement of RBCs due to hypoxic conditions, disrupted water balance (osmotic stress), or macrocytic anemia in fish subjected to stress (Tort *et al.*, 1988). The reduction in MCV following brief exposure, along with diminished hemoglobin levels, suggests that the red blood cells have contracted, either due to hypoxia or microcytic anaemia. At this juncture, microcytosis may result from a reduction in haematocrit during exposure (Adeyemo, 2005). The MCH value was recorded in this investigation for both control and infected *Channa striatus* and *Channa punctatus* fishes. The reduced MCH value in EUS-infected fish indicates hypochromic anemia. The current results align with the research conducted by Benarjee *et al.* (1986) on the toxicological effects of organophosphorus insecticides on the haematological and histological alterations in *Clarias batrachus*. A notable disparity in MCHC was observed between the control and infected fish.

The MCHC, a reliable predictor of RBC enlargement, is unaffected by blood volume or cell count, which may lead to misinterpretation when new cells with varying hemoglobin concentrations enter the bloodstream (Soivio *et al.*, 1981). Nonetheless, MCH and MCHC exhibited a significant rise in week 4 among groups administered with L. inermis soluble fractions (Venkatachalam *et al.*, 2014). The elevation in MCHC value is attributed to macrocytic anemia, and a high MCHC level signifies an increased hemoglobin concentration per unit of red blood cells (Robbins *et al.*, 1974). MCV, MCH, and MCHC were computed indirectly based on RBC, Hct, and Hb; hence, their variations are directly associated with these hematological parameters (Kandeepan, 2012).

The current study demonstrated a significant reduction in the hematological parameters in *Channa striatus* and *Channa punctatus* during various bacterial and fungal inoculations. The decrease was observed in both the fish infected by *Aeromonas hydropila, Staphylococcus aureus, Pseudomonas aeruginosa*, and *Salmonella salmonicida*, as well as in those affected by *Ascomycota species such as Aspergillus flavus, Fusarium solani, Rhizophus stolenifer, Aspergillus fumigatus, Penicillium chrysogenum, Aspergillus niger*, and *Trichoderma viridae*. The selected fish were sourced from three lakes. Hasanparthy, Dharmasagar, and Bandham were afflicted by Epizootic Ulcerative Syndrome (EUS). The alteration in haematological parameters may result from EUS infection, seasonal variations, and the biological habits and habitats of the fish. This study indicates that fish are highly sensitive to alterations in their ecology, which is indicative of their haematological problems.

This study successfully demonstrates the induction of infection in EUS in *Channa striatus* and *Channa punctatus*, indicating the infectious nature of the condition. The susceptibility of Channa species to EUS has considerable ramifications for aquaculturists in the southeastern area of India. Considering the significant transmission potential, polyculture involving Indian main carps, *Channa striatus* and *Channa punctata*. The potential application of haematological parameters as a resource in aquaculture and fisheries management.

CHAPTER 3

Histopathology

INTRODUCTION

This chapter provides a comprehensive overview of the common histological changes observed in fish afflicted by EUS, referencing specific species observations. The initial skin lesions in certain samples have been identified as mostly regions of epithelial necrosis accompanied by adjacent edema, hemorrhaging of the underlying dermis, and some infiltration of inflammatory cells. Confirmation of fungal involvement in the majority of these early samples has not been feasible. Nevertheless, a limited number have included a minor quantity of hyphae. Fungal hyphae were identified in the epidermis of certain early -stage infected fish from India (Viswanath *et al*., 1997). Roberts *et al.* (1989) similarly investigated early lesions during an EUS outbreak in a captive population of Indian large carp. They noted an acute necrotizing myopathy, more severe than often found in wild fish, disseminated over a broad region beneath the active skin lesion. The epidermis near the ulcer borders was degenerated and thickened, including a minimal quantity of fungal hyphae encased within an epithelioid capsule. The dermal blood vessels exhibited significant hyperaemia, with some surrounded by a sheath of lymphoid or myeloid cells, potentially linked to viral infection, despite the absence of detectable viral inclusion bodies, and subsequent clinical changes.

All infected fish species exhibit significant degenerative changes in skin and muscle tissue, with minimal disruption of internal organs. Advanced lesions have widespread necrotizing granulomatous mycosis of the underlying muscle fibers, marked by distinctive branching aseptate, invasive fungal mycelium. A considerable amount of germs may be present on the surface of specific advanced lesions. As the lesion progresses, fungal cells are progressively enveloped by dense layers of host epithelioid cells, and specific areas may show signs of myophagia and healing. In specific advanced lesions, fungal hyphae may infiltrate the abdominal viscera, possibly acting as the primary cause of death. A substantial number of mycotic granulomas have been detected in the kidneys, livers, and digestive tracts of many fish species, including spiny eels, Cirrhinus mrigal, Colisa lalia, Channa sp., Puntius sp., Esomus sp., Mugil sp., Valamugil sp., Therapon sp., and Glossogobius sp. Sillago species. (Chinabut, 1990; Ahmed &

Podeti Koteshwar Rao

Hoque, 1998; Wada *et al.*, 1994; Viswanath *et al.*, 1998). Wada *et al.* (1994) identified mycotic granulomas in the abdominal adipose tissue, pancreas, gonads, spleen, central nervous system, and heart of dwarf gourami. Vishwanath *et al.* (1998) subsequently demonstrated fungal invasion of the esophagus and spinal cord of mullet, as well as intermuscular bones of Puntius. The internal organs of ill fish, except those impacted by fungal hyphae, display minimal histological changes, which Roberts *et al.* (1986) observed may occasionally arise from background pathology. Palisoc (1990) observed minimal tissue damage in the kidneys of striped snakeheads, marked by an increased presence of melanomacrophage centers, haemosiderin pigments, and a restricted quantity of mitotic figures. Regions of the spleen demonstrated a considerable enhancement in white pulp formation, but the heart, liver, and gills showed minimal histological modifications. Still, no abnormalities were observed in the stomach and intestine. Supplementary kidney samples have revealed tubular vacuolar degeneration, along with granular obstruction and degeneration or localized proliferation of hematopoietic tissue.

Chinabut (1990) determined that these traits were consistently more apparent in armed spiny eels. Pancreatic specimens sometimes display acinar necrosis (Callinan *et al.*, 1989) along with eosinophil and inflammatory cell infiltration. Prior to bacterial or fungal involvement, mild localized hepatic cellular degeneration may manifest in the liver. The only consistent haematological change in affected fish is a significantly diminished hemoglobin concentration resulting from extra- and intra-vascular erythrocyte deterioration (Tangtrongpiros *et al.*, 1985).

THE MATERIALS INCLUDING TECHNIQUES SAMPLE SELECTION

Channa striatus and Channa punctatus exhibiting visible indications of microbial and mycotic infection were obtained from Hasanparthy, Dharmasagar, and Bandham lakes in Warangal district, Telangana, India. The specimens (fish) were gathered alive using a fishing net and promptly transported to the laboratory for additional clinical assessment in plastic containers filled with oxygenated water.

The specimens were profoundly anesthetized by immersion in a 5 ml/L water solution of ethylene glycol monophenyl ether. Following the dissection of the fish, tissue samples from the skin, gills, liver, and pancreas were meticulously extracted and preserved in fixatives. Skin, gill, liver, and pancreatic tissues were preserved in Bouin's, Zenker's, Susa, Carnoy, and Formal Calcium standard fixatives according to established techniques. The diseased fish exhibited red blotches on their bodies, excessive mucus production, damage, and lethargy. The microbial examination was conducted on various infected tissues obtained from

Channa striatus and Channa punctatus, specifically the skin, gills, liver, and pancreas. The bacterial species include Aeromonas hydropila, Staphylococcus aureus, Pseudomonas aeruginosa, and Salmonella salmonicida. The fungal germs identified in the aforementioned organs include Aspergillus flavus, Fusarium solani, Rhizopus stolonifer, Aspergillus fumigatus, Penicillium chrysogenum, Aspergillus niger, and Trichoderma viride.

PROCESSING FOR HISTOPATHOLOGICAL INVESTIGATIONS

Samples of skin, gills, liver, and pancreatic tissues were extracted from the fixative.

Hydration

The tissues were washed with saline solutions according to established protocols.

Dehydration

The process involved placing the tissue in sufficient 70% alcohol, then transferring it to 90% alcohol for 1 change and 2 changes of 100% alcohol.

Clearing

Tissues were cleared using a clearing agent, specifically xylene.

Embedding and Impregnating

The impregnation agent employed was paraffin wax, which enabled effortless entry into the tissue without inducing structural damage or significant shrinkage or crystal formation.

Trimming

The embedding block containing tissue was meticulously clipped to ensure that just a single tissue specimen was prepared for sectioning.

Mounting of the Block

An iron block measuring 30 mm was utilized to secure the block at the appropriate angle and position for cutting.

Section Cutting

The tissue block was sectioned to a thickness of 5μm using a microtome.

Floating and Mounting of Sections

The cutting sections were permitted to float in a water bath to prevent creases. Moreover, it was observed that the sliced parts were securely affixed to the slides and positioned correctly. The slide received an egg albumin coating prior to the application of the sections.

Staining the slides with Azan and mounting

The stained sections were examined under a light microscope, and microphotographs were captured for pathological analysis.

RESULTS

In this study, pathological alterations were noted in the skin, gills, liver, and pancreas of Channa striatus and Channa punctatus due to infections caused by *Aeromonas hydrophila, Staphylococcus aureus, Pseudomonas aeruginosa, Salmonella salmonicida, Aspergillus flavus, Fusarium solani, Rhizopus stolenifer, Aspergillus fumigatus, Penicillium chrysogenum, Aspergillus niger,* and *Trichoderma viride.*

Indeed, in numerous instances, their presence accounts for some variations in the inflammatory response. Fish infected by EUS have exhibited significant alterations due to the infecting agents. Nonetheless, the predominant finding was bacterial-associated necrosis, and the ramifications of these lesions should likely guide inquiries into the health determinants of these fish.

CLINICAL AND PHYSICAL INDICATORS OF INFECTED FISH

The current study indicates that the infected fish exhibit the following gross observations: dullness, loss of balance, anorexia, sluggish movements, surface swimming, lethargy, and irregular, spiral swimming patterns. Coarse, hyperpigmented skin hemorrhage at the fin bases, occasionally accompanied by skin ulcers, and unilateral or bilateral exophthalmia. The protrusion of the eyeball, preventing the eyelids from covering it, isdue to ocular opacities with hemorrhages, and abdominal distension. Lilley *et al.* (1998) indicate that infected fish exhibit reduced appetite and lethargy, often floating just below the surface or swimming with their heads above the water. The current study suggests the emergence of pinhead-sized red spots on the body surface, head, fins, caudal peduncle, dorsum, or operculum, without any observable hemorrhages or ulcers. The intermediate stage lesions are characterized by tiny (2-4 cm) cutaneous ulcers accompanied by scale loss, hemorrhages, and edema. The advanced lesions manifest on many regions of the fish's body, developing into extensive necrotic open sores, ultimately leading to death.

Particular affected species, such as the striped snakehead, can endure even more severe, chronic lesions that may have entirely obliterated the caudal peduncle or eroded into the brain or abdominal cavity, occasionally exposing the swim bladder. The liver may exhibit pallor or a green hue, accompanied by hepatic necrosis and the presence of fat droplets. These are ostensibly compromised by bacterial toxins, resulting in a loss of structural integrity. The primary lesion consisted of multifocal skin and liver lesions, characterized by severe, acute-t--subacute, extensively necrotic regions.

HISTOPATHOLOGY OF DERMIS

Histologically, diseased fish frequently have epidermal degradation due to colonization by Aeromonas species, with minimal accompanying inflammation. In Channa striatus and Channa punctatus, extensive epidermal loss has been observed across significant areas of the body, and the infection has been recurrently diagnosed since then. Histopathological alterations in skin tissue infected with S. parasitica exhibited epidermal loss and necrotized hypodermis, according to Miyazaki and Egusa (1972).

The muscle layer exhibits inflammation characterized by necrotic tissue debris and the presence of many well-formed layered granulomas encircled by hyphae. Hatai (1980) Granulomas exhibited fibrillar features resulting from hyphal infection. Tissue injected with A. laevis exhibited symptoms comparable to those observed with S. parasitica. Necrotic epidermis and hypodermis are accompanied by a granulomatous reaction in the muscle. Histological examination of tissue injected with A. niger revealed the absence of granuloma formation in the musculature, with no hyphae detected in the deeper layers; however, degeneration of the epidermis and edema in the underlying hypodermis and musculature were observed (Hatai *et al.*, 1994; Hussian *et al.*, 2013). A granulomatous reaction resulting from Aphanomyces infection has also been documented by Qureshi *et al.* (2001), Sosa *et al.* (2007), and Chauhan and Qureshi (2012).

Histopathological examinations of tissue infected with A. niger revealed no granuloma formation; nevertheless, deteriorated muscle accompanied by necrotized hypodermis and epidermis was seen (Rekha Chauhan *et al.*, 2014). Histopathological examinations of the skin revealed epithelial desquamation, resulting in erosion and ultimately leading to ulceration in the infected region. There was a varying extent of damage in the dermis and hypodermis. Epidermal and dermal cells exhibited vacuolar degeneration and localized necrosis. The dermis and hypodermis exhibited necrosis with fragments of fungal hyphae alongside localized aggregations of melanomacrophages. Histopathological analysis of the diseased skin and musculature of Channa maurilus revealed diverse forms of tissue damage, destruction of the epidermal layer, accompanied by total necrosis of the dermis and hypodermis. Fungal hyphae were distinctly identified in the muscle layer (Rekha Chauhan *et al.*, 2014). Histopathological manifestations of mycotic infections have been previously documented (Laxmareddy *et al.*, 2013; Hussian *et al.*, 2013; Chauhan *et al.*, 2014a), with Aspergillus sp. causing significant, dose-dependent tissue destruction.The reports of the aforementioned researchers corroborate the infection noted in the current study. The inflammatory response primarily reflected a host response to germs and fungi. The skin subtly protruded above the surface. The muscle fibers exhibited vacuolation, granulomatous inflammation, necrotic tissue debris, and the presence of inflammatory cells. Fungal hyphae of varying lengths were detected in muscle tissue. The granulomas encircled hyphae that occupied the necrotic regions of longitudinal and circular muscles. The skin exhibited localized epidermal sloughing accompanied by hyperplasia of alarm and mucous cells. The divided dermis was nearly united by fibrous connective tissues aligned parallel to the epidermal layer. The quantity of dead cells resulting from sarcolysis diminished. Active macrophages were identified alongside localized lymphocyte aggregation in the healing area. Congested and hemorrhagic dermis exhibited excessive accumulation of lymphocytes and melanomacrophages, together with hyaline changes in the epidermal basement membrane. Infected fish displayed full mucosal erosion, with regrowth of muscle fibers. The regenerated muscular bundle is nearly supplanted by fibrotic tissue, exhibiting significant degenerative alterations in the muscles and necrotic changes in the muscle fibers. The number of necrotic muscle fibers increased, accompanied by swelling of the myoseptum and sarcolysis, resulting in an enlarged lesion area. Hemorrhage was noted within the necrotic region, and a limited number of melanin-containing cells began to appear beneath the basement membrane of the epidermis. The injured segment of the dermis remains unattached. Musculature was distinctly seen, with the muscle fibers exhibiting hyalinization and significant nuclear damage. The tissue response was less prevalent when the activation of melanophores was disregarded. The histopathology of gills reveals significant pathological alterations in the architectural structure of fish gills due to EUS infection. The modifications encompass Epithelial Lifting (EL), bulging of primary gill filament tips (BTPG), degeneration of secondary lamellae (DGSL), curling of secondary gill filaments (CSG), atrophy of secondary lamellae (ASL), and fusion of secondary gill filaments (FSG).

The damage to the gills of fish with elevated levels of EUS infection was substantial. There was significant shortening and clubbing of the ends of the secondary gill lamellae, fusing of adjacent secondary gill lamellae, and necrosis in the primary lamellae. Nuclear hyperplasia and enlargement were also seen. In addition to these alterations, pyknotic nuclei, vacuolization, degeneration of epithelial and pillar cells, and detachment of the epithelial layer from the secondary lamellae were also noted. Bacterial gill illness is the most fatal among these types and is discussed below. The current investigation found hyperplasia and hypertrophy of epithelial cells, epithelial lifting, lamellar disorganization, lamellar aneurysm, rupture of the lamellar epithelium, rupture of pillar cells, and necrosis resulting from EUS infection. Many researchers have discovered analogous pathogenic alterations related to bacterial and fungal poisoning. De Silva and Samayawardhena (2002) noted the abnormal morphology of gill lamellae, heightened vacuolation in epithelial cells, lamellar fusion, and total degradation of gill lamellae in Poecilia reticulata exposed to chlorpyrifos. The damage manifested in the secondary gill lamellae, characterized by mild mucous deposition and exfoliated nuclei, along with the fragmentation of muscle fibers in freshwater fish subjected to acute and chronic EUS infections. Bacteria and fungi impact the organ systems, particularly during the winter season. It can be posited that pathological changes such as epithelial cell hyperplasia, epithelial lifting, and lamellar fusion may augment the contact area between toxicants and the gill's vascular system, leading to respiratory impairment and detrimental effects on fish health. Severe hyperplasia leads to the fusing of secondary lamellae and often causes abnormalities, including blood congestion, hypertrophy of epithelial cells, and lamellar disarray. Marina *et al.* (2007). Hyperplasia of epithelial cells, subsequent lamellar fusion, goblet cell proliferation, and the migration of Eosinophilic Granular Cells (EGCs) to the gills of fish infected with these parasites have been documented (Mehdi Raissy and Mahsa Ansari, 2011). Rekha Chauhan *et al.* (2014) reported that gill lamellae had damaged epithelium and fungal hyphae encased by many layers of fused secondary lamellae, indicative of a severe infection. During October and November, the primary and secondary gill lamellae were systematically organized, and no notable pathogenic symptoms were detected in the gill structure. In December, the primary gill lamellae exhibited hypertrophy, and a limited number of blood cells were present at the base of the secondary lamellae. Inflammatory cells and minor hemorrhages were seen in the primary gill lamellae across all farms throughout this winter season. In January, the primary gill lamellae were significantly impacted, resulting in pronounced hypertrophy and hyperplasia, with portions of the secondary gill lamellae absent (Chandra *et al.*, 2012).

HISTOPATHOLOGY OF THE LIVER

Bacterial, fungal, environmental pollution, and reduced oxygen levels trigger EUS infection, resulting in distinct pathological alterations in the liver tissue of the fish species *Channa striatus* and *Channa punctatus*. Histopathological analysis of the livers of *C. striatus* and *C. punctatus* infected with bacteria and fungi revealed ruptures between hepatic cords, delineation of the cell membranes of hepatic cells and connective tissues, extensive blood spaces, and complete separation of blood vessels from hepatic cells in certain areas, along with enlarged hepatic and bile ducts. Infectious liver exhibits the creation of numerous minute voids between the hepatic cords, total loss of connective tissues in certain areas, evident atrophy of hepatic cell nuclei, and homogeneity throughout the hepatic mass.

The alterations encompass degraded hepato-pancreatic tissue (DGHP), blood cells interspersed among hepatocytes (BC), the emergence of blood streaks within hepatocytes (ABS), the formation of vacuoles (FV), as well as atrophy, necrosis, and modifications to hepatocytic cell walls and hepatic cord arrangement. The degenerative alterations are exacerbated in chronic illness states. The organ is primarily linked to detoxification and biotransformation processes in the liver, attributable to its function, location, and blood supply (Vander Oost *et al.*, 2003). Rodrigues and Fanta (1998) posited that fish liver histology could serve as a model for examining the relationships between environmental influences and hepatic structural functions. Environmental factors encompass biotoxins, parasites, pathogenic microorganisms, physicochemical characteristics, and pollutants such as pesticides, hydrocarbons, polychlorinated biphenyls (PCBs), and heavy metals (Brusle and Anadon, 1996). The predominant lesions in the diseased liver included vacuolar degeneration of hepatocytes, focal necrotic regions, and clusters of inflammatory cells interspersed among the hepatocytes. Dilatation and thrombosis in central veins, together with dilatation and congestion in blood sinusoids and intravascular hemolysis in hepatic and hepatoportal blood arteries, were also noted. Furthermore, distinct regions of coagulative necrosis and fibrosis were clearly observed. The organ primarily impacted by Aeromonas is the liver, which is the predominant site of cellular degeneration (Vanderoost *et al.*, 2003). Vascular dilatation, intravascular hemolysis, and thrombus development in blood arteries, leading to blood stasis, may contribute to cellular deterioration and necrosis in the liver (Prema Raju, 2011). In October and November, the hepatocytes of the liver were typically organized. No notable pathological differences were observed in the liver's structure. In December, necrotic hepatocytes exhibiting many pyknotic nuclei were seen. In January, hepatocytes were significantly compromised and ruptured, resulting in necrotic hepatocytes containing vacuoles, alongside the presence of blood cells and inflammatory cells (Chandra *et al.*, 2012).

The current investigation identified significant abnormalities in the livers of EUS-infected *C. striatus* and *C. punctatus*, including irregular nuclear morphology, nuclear hypertrophy, nuclear vacuolation, and the presence of eosinophilic granules in the cytoplasm. Bile stagnation is characterized by the presence of brownish-yellow granules inside the cytoplasm. The cytoplasmic and nuclear degraded cells are prevalent. Melanomacrophages were recognized as spherical cell aggregates containing dark yellowish granules of varying sizes, typically located near the arteries. The histological examination of the pancreas in the control *Channa striatus* and *Channa punctatus* revealed a pancreatic mass localized around the branches of the hepatic portal veins, infiltrating the hepatic cells. The gap is observable between hepatic and pancreatic cells. The pancreatic mass comprises two components: the exocrine and endocrine portions. Exocrine cells are bigger and elongated, whereas endocrine cells are smaller and spherical. Exocrine cells are organized in an acinar structure with prominent nuclei, whereas endocrine cells are dispersed among the hepatic portal veins and the exocrine pancreas. The endocrine components of pancreatic tissue include beta cells, which are round and indistinct, whereas alpha cells are smaller and more pronounced. Blood arteries and hematologic spaces exhibit conspicuous erythrocytes.

The subsequent necrotic cells exhibited pyknotic, karyorrhectic, karyolytic nuclei, or were devoid of nuclei. Significant fatty degeneration was noted in the liver of fish infected with S. dysgalactiae. Additionally, localized regions of mononuclear inflammatory cell infiltration were observed in the hepatic tissue of fish infected with (*E. gallinarum, E. fecalis*, and *S. dysgalactiae*). The previous infiltrates were noted in the liver infected with *E. fecalis* and *S. dysgalactiae*, replacing numerous focal regions of necrotic and entirely destroyed hepatocytes. The hepatopancreas exhibited vacuolation of the acinar cells and dispersed necrotic cells, some of which displayed pyknotic nuclei or were devoid of nuclei. The hepatopancreas of fish infected with *E. fecalis, S. anginosus*, and *L. garvieae* exhibited entirely disordered and destroyed acini, along with infiltration of inflammatory cells among the acinar cells. Moderate infiltration of inflammatory cells was found in the portal triads of fish infected with *S. anginosus*. The augmentation of melanomacrophage centers, characterized by clumps of pigmented cells, was a notable finding across all fish species (Nehal *et al.*, 2014).

HISTOPATHOLOGICAL ANALYSIS OF THE PANCREAS

In freshwater fish infected with bacteria and fungi, pancreatic tissue was diminished, with indistinct exocrine and endocrine components, potentially due to pancreatic necrosis, disrupted junctions between hepatic and pancreatic tissues, and the presence of clear vacuoles. Ajit *et al.* (2014) indicate that pancreatic tissues are atrophied and diminished, leading to a loss of distinction between the exocrine and endocrine components. At this junction of hepatic and pancreatic tissue, distinct vacuolization is noted, possibly due to hepatic necrosis. The examination of the diseased pancreas revealed degeneration of certain glandular cells and the presence of a few lymphocytes surrounding the acini. No granules were detected, and significant tissue damage was noted in the liver tissues. Liver tissues exhibited localized to coalescing necrosis of the interstitial tissue. In the liver, sinusoids were dilated and partially occupied by granular, eosinophilic substance (edema fluid). A multitude of cells with pycnotic nuclei was seen randomly dispersed throughout the hepatic tissue. The histological alterations found in the liver included cytoplasmic degeneration in hepatocytes, atrophy, vacuole development, blood vessel rupture, and disorganization of hepatic cords (Prema Raju, 2011).

DELIBERATION

The freshwater fish samples analyzed in this study were found to be contaminated with bacteria and fungi. The acute fins of Channa species can readily lacerate the skin of nearby fish, enabling the facile ingress of bacteria and fungi. The histological changes are present in the infected tissues of the skin, gills, liver, and pancreas of *Channa striatus* and *Channa punctatus*.

In recent decades, the aquaculture industry has transformed fish production, characterized by a tendency towards intensification through the use of recirculating systems (Larkin and Sylvia, 1999). The primary motivation was to minimize environmental impact, enhance management of the rearing environment, and improve biosecurity measures. Nevertheless, the enhanced management of raising circumstances has prompted the industry to augment the stocking densities of target fish. This has led to a rise in disease outbreaks, characterized by accelerated transmission and heightened mortality rates (Bondad-Reantaso *et al.*, 2005; Whittington and Chong, 2007; Peeler *et al.*, 2010; Gonçalves and Gagnon, 2011).

Insufficient epidemiological data regarding fungal pathogens affecting fish currently exist. The recent emergence of *S. destruens*, which threatens European fish diversity (Gozlan *et al.*, 2005; Andreou *et al.*, 2012), suggests that ecological impacts akin to those observed in chytrid parasites affecting amphibians may be occurring in freshwater fish populations (Gozlan, 2012). The vast majority of *S. destruens* infections are caused by the invasion of a healthy fish host. Spikmans *et al.* (2013) assert that further monitoring of invading wild fish communities would reveal the presence of this fungal-like infectious pathogen. Indeed, there is insufficient data on fungal pathogens, particularly for fish from wild populations. The predominant documentation of fish fungal and fungal-like infections has primarily originated from the aquaculture sector, with scant studies concerning fungal development in wild fish populations. The primary reason is probably a combination of the absence of external pathogenic specialization in infected hosts and the chronic character of certain diseases produced by these fungal and fungal-like infections, which, unlike viral pathogens, propagate over extended durations.

CONCLUSION

The Epizootic Ulcerative Syndrome (EUS) presents a considerable threat to global aquaculture, impacting numerous fish species. The effective management of EUS depends on a synthesis of precise diagnostic methods, preventive actions, and control strategies. Advanced diagnostic instruments, such as molecular methods like PCR, along with histopathological and histochemical analyses, have improved the early detection and surveillance of the disease. Preventive approaches, including enhancing water quality, regulating environmental stressors, and implementing biosecurity standards, are crucial in reducing epidemics. Molecular diagnostics and environmental monitoring are crucial for prompt response. Preventive interventions, such as water quality control, biosecurity measures, and stress alleviation, are essential in diminishing vulnerability to the disease. Control strategies, encompassing antifungal therapies, vaccine development, and the amalgamation of conventional and contemporary management techniques, are crucial for alleviating the effects of EUS. Ongoing research and collaboration among stakeholders, including fish farmers, researchers, and policymakers, are essential for promoting sustainable aquaculture methods. Through the implementation of comprehensive and proactive methods, the aquaculture sector can effectively address EUS, safeguarding fish health, minimizing economic losses, and fostering environmental sustainability. Continued investigation into disease processes, host-pathogen interactions, and innovative therapeutic strategies will facilitate enduring solutions to EUS. Epizootic Ulcerative Syndrome (EUS) remains a considerable menace to aquaculture, resulting in major economic detriment and impacting aquatic biodiversity. Effective management of EUS necessitates a comprehensive approach that includes advanced diagnostic methods, strong prevention strategies, and dependable control measures.

AI and ML-driven preventive techniques represent a significant transformation in aquaculture disease management. Predictive modeling facilitates the early detection of outbreak hazards, whereas optimization methods guarantee sustainable and efficient agricultural operations. Decision support systems facilitate equitable access to sophisticated technology, enabling farmers and aquaculture managers to proactively address EUS hazards. These innovations collectively improve resilience, productivity, and sustainability in global aquaculture systems. The management of Epizootic Ulcerative Syndrome (EUS) in aquaculture is on the verge of transformation through the implementation of Artificial Intelligence (AI) and Machine Learning (ML). These technologies provide robust solutions for the issues associated with disease detection, prevention, and control, hence maintaining the sustainability of aquaculture activities.

Revolutionizing EUS Management with AI and ML

Artificial Intelligence and Machine Learning facilitate a transformative change in Endoscopic Ultrasound management by implementing:

Improved Diagnostic Precision

AI-driven picture recognition and predictive models provide swift and dependable disease identification, exceeding conventional techniques in both speed and accuracy. Data-Driven Prevention, predictive analytics, and decision-support systems provide proactive interventions utilizing environmental and historical data, hence mitigating outbreak risks. Efficient Control Strategies: AI-driven treatment recommendation systems, together with continuous monitoring, enhance interventions while reducing resource consumption and environmental impact. These inventions enhance fish health and farm productivity while also addressing global issues such as food security and sustainable aquaculture. The significance of technology adoption: To harness the potential of AI and ML, it is essential to surmount obstacles to adoption and cultivate a collaborative ecosystem. Empowering small-scale farmers through accessible, cost-effective technologies tailored to their needs can democratize the advantages of AI, facilitating widespread adoption. Advancing Sustainability: AI technologies enhance resource efficiency, diminish waste, and lessen dependence on detrimental methods, harmonizing aquaculture with environmental conservation objectives, enhancing Resilience by incorporating real-time monitoring, early warning systems, and sophisticated analytics. Aquaculture operations may more effectively respond to problems such as climate change and disease outbreaks. The integration of AI and ML is essential for the future of sustainable aquaculture. By embracing these technologies, governments, researchers, technology developers, and farmers can advance the aquaculture industry towards a future where EUS and other illnesses are effectively managed, safeguarding the health of aquatic ecosystems and the lives of millions reliant on them. Artificial Intelligence and Machine Learning provide the means to harmonize productivity with sustainability, facilitating a robust and thriving aquaculture industry.

CHAPTER 4

Histochemical Analysis

INTRODUCTION

Histochemistry has significantly advanced the comprehension of biological phenomena and clinical medicine. Histochemical techniques facilitate the analysis of the localization of proteins, lipids, glycogen, and other substances, as well as molecular alterations at the cellular level. The primary benefit of histochemistry is its ability to analyze biological events within specific cells. The modified condition of cellular components can be examined by histochemical methods. The histochemical assays indicate the location of chemical products resulting from cellular activity. This work examines the histochemical characteristics of the skin, gills, liver, and pancreas of *Channa striatus* and *Channa punctatus* to analyze the differences between control samples and those affected by bacterial and fungal infections.

The skin serves as the principal barrier against environmental factors, facilitating proper internal physiological functions (Sarasquetel *et al.*, 1998). This condition is significant in numerous disease processes. The skin mucus of fish may play a crucial role in natural defense against parasites and harmful microorganisms (Fletcher, 1978), in addition to potentially serving osmoregulatory and lubricating functions (Van Oosten, 1957; Rosen; Conford, 1971). The defense may be mechanical, resulting from the constant generation of mucus (Pickering, 1974). The presence of lysozymes and IgM complement components indicates that mucus plays an active role in the immune system (Hjelmeland *et al.*, 1983).

This encompasses a range of specific interests, from architecture to histochemistry and necrotic diseases of the skin, particularly its zoonotic significance. The structural integrity and histochemical intricacies are essential for the skin's reliance in this regard. Despite considerable uniformity in the fundamental organization of vertebrates, particularly regarding the dermis, epidermis, and basement membrane, light and electron microscopy have elucidated the structural architecture of the skin and its protective functions for the organism.

Most prior research in histochemical investigations of skin, such as Askawa (1970), examined the histochemistry of mucus on the epidermis of the eel

Podeti Koteshwar Rao

Anguilla japonica. Mittal and Munshi (1971) conducted a comparative research on the skin structure of specific air-breathing freshwater teleosts. Harris *et al.* (1973) conducted a histochemical examination of mucous cells in the epidermis of the brown trout, Salmo trutta. Banerjee and Mittal (1975) conducted research on the histochemistry and functional organization of the skin of the "livefish" *Clarias batrachus*. Mittal and Banerjee (1975) conducted investigations on histochemistry and the structural composition of the skin of the murrel, Channa striata. Mittal *et al.* (1976) investigated the protein and carbohydrate histochemistry concerning keratinization in the epidermis of Barbus sophor (Cyprinidae, Pisces). Mittal and Whitear (1979) documented the keratinization of fish skin, specifically focusing on the catfish Bagarius bagarius. Mittal *et al.* (1980) delineated the fine structure and histochemistry of the epidermis of the fish Monopterus cuchia, whereas Hjelmeland *et al.* (1983) examined skin mucus protease from the rainbow trout, Salmo gairdneri Richardson, and its biological importance.

Sultana and Rao (1990) reported observations regarding the structure and histochemistry of the skin of *Mystus gulio*. Mittal *et al.* (1994) conducted a histochemical investigation of glycoproteins in the unicellular glands of the epidermis of the Indian freshwater fish Mastacembelus panculus. The histochemical examination of glycoproteins in the unicellular glands of the epidermis of the Indian freshwater fish Mastacembelus. Mittal *et al.* (1995) described the surface architecture of the skin of the Indian catfish, Bagarius bagarius. Gonzalez *et al.* (1996) examined the histological and histochemical properties of lymphocytosis in seabream, *Sparus aurata*. Sarasquetel *et al.* (1998) examined the histochemical aspects of lymphocystis illness in the skin of gilthead seabream, *Sparus aurata*. Park and Kim (1999) documented the structure and histochemistry of the skin of the mud loach, Misgurnus anguillicaudatus.

Park *et al.* (2000) conducted a histological examination of the skin of the amphibious fish, Periophthalmus modestus. Park *et al.* (2001) investigated the morphology and histochemistry of the skin of the mud loach, Misgurnus mizolepis, concerning cutaneous respiration. Zaccone *et al.* (2001) elucidated the structural, histochemical, and functional characteristics of fish epidermis. Saikia and Rai (2002) conducted histochemical research on the functional organization, amounts, and distribution of various carbohydrate elements in the skin club cells of the catfish Heteropneustes fossilis (Bloch) (Heteropneustidae, Pisces) under acid stress. Park (2002) examined the integumentary anatomy of the air-breathing mudskipper fish, Periophthalmus magnuspinnatus. Jong-Young Park (2002) investigated the morphology and histochemistry of the skin of the Korean spined loach, Iksookimia koreensis (Cobitidae), concerning breathing. Park (2002a) examined the morphology and histochemistry of the skin of the spined cobitid fish, Iksookimia koreensis, concerning respiration. Park (2002b) studied the structure of the skin of the air-breathing mudskipper fish, Periophthalmus magnuspinnatus. Park *et al.* (2003) contributed to the understanding of the structure and histochemistry of the skin of the torrent catfish, Liobagrus medtadiposalis. Swati Mittal *et al.* (2004) conducted a scanning electron microscopy and histochemical study on the operculum of the peppered loach, Lepidocephalichthys guntea.

The epidermis is separated into three layers: an uppermost layer, a middle layer, and a basal layer. The outermost layer consists of multiple strata of epithelial cells. Chloride cells, outermost cells, mucous cells, and sacciform granular cells are located at the surface. The intermediate layer consists of unicellular glands, mucous cells, and sacciform granular cells. Epithelial cells occupy the interstitial areas between the glandular cells (Sunita *et al.*, 2004).

Dutta and Rai (2005) conducted studies on lipids in the skin of the catfish *Clarias batrachus* and described the presence of proteins and nucleic acids in its skin. Guerra *et al.* (2006) characterized the stratum adiposum, a distinctive structure of the African catfish skin (*Clarias gariepinus*). Rai and Saikia (2007) examined the impact of acid stress on the histochemical response of alkaline phosphatase activity in the skin cells of the catfish Heteropneustes fossilis. Rai and Saikia (2008) observed the effects of acid stress on the functional organization and distribution of carbohydrate constituents in the skin of Heteropneustes fossilis. Bonilla *et al.* (2008) analyzed the skin histology and morphometry of the fish Eremophilus mutisii. Tarun *et al.* (2010) detailed the histochemistry and functional organization of the dorsal skin of *Andstrus ddlichopterus*. Mohamed *et al.* (2010) examined the effects of experimental infection with Gyrodactylus species on the density of skin mucus in catfish fry (*Clarias gariepinus*), focusing on the pathological alterations observed in the skin histochemistry of freshwater catfish *Clarias batrachus* (Laxma Reddy Benarjee, 2011).

The fish skin is the protective structure that envelops the body, safeguarding it from the infiltration of bacterial and fungal infections or allergens, as well as preventing the loss of water, solutes, or minerals. The outside-in and inside-out barrier functions rely on the epidermis, a stratified cellular layer. Mucus envelops the epidermis in fish and amphibian tadpoles, whereas distinct cornified cellular layers (stratum corneum) form the outermost epidermal barrier in adult amphibians, reptiles, birds, and mammals. Kubo (2011) states that the skin of teleosts is particularly distinctive and histologically diversified (Fast *et al.*, 2002). It markedly differs from that of mammals, as it secretes mucus that plays a role in immunological processes (Salinas *et al.*, 2011). The shape and function of the organism demonstrate its adaptability to the physical, chemical, and biological characteristics of the aquatic environment, as well as its evolutionary history. The water environment harbors several harmful organisms. According to Magnadottir (2010), the integument of aquatic vertebrates serves as a critical first line of defense against environmental infections, and it is equally vital for the respiratory and digestive systems. Due to the close interaction of fish with their environment, skin illnesses are more prevalent in fish than in terrestrial vertebrates and represent a main health concern for aquatic animal practitioners (Groff, 2001). It is a substantial organ that is continuous with the linings of all bodily orifices and also envelops the fins. Besides functioning as a mechanical barrier, it constitutes a metabolically active tissue (Bullock *et al.*, 1974). Fish integument is a multipurpose organ, with its components playing significant roles in protection, communication, sensory perception, movement, breathing, and ion management. According to Marshall *et al.* (2010), excretion and heat regulation are facilitated by the intricate structure and cellular makeup of the skin (Elliott, 2000). These functions (primarily immunity, osmoregulation, respiration, and excretion) are particularly crucial in fish larvae due to the reliance on the skin during early developmental stages, which is attributed to the high surface-to-volume ratio that diminishes as development progresses (Varsamos *et al.*, 2005). The integument, or skin, serves as the barrier that both isolates and safeguards a fish from its surroundings, while simultaneously facilitating the majority of interactions with the external environment (Maria, 2012).

The pathogenicity of *Aspergillus fumigatus* and *Aspergillus niger* has been documented in freshwater fish (Chauhan, 2013). Adverse environmental conditions and low temperatures will elevate fungal infections in freshwater and marine fish. Infections of *L. Rohita* and silver carp, Hypophthalmichthys molitrix, have also been observed to be associated with Aspergillus sp. and Alternaria sp. (Iqbal and Ramsha, 2013). This type of pollution has been identified as a cause of EUS infection, which is a major disease in India and throughout South Asia. Commercially significant fish populations are diminishing due to overfishing, water pollution, and insufficient aeration in lakes. The impact of these illnesses on species in freshwater habitats, other researchers have isolated *Aspergillus* sp. from freshwater fish (Chauhan, 2014a). Fishermen in India are encountering significant challenges due to fungal illnesses. Fungi are most affected by temperature fluctuations and contaminated water conditions, which promote the proliferation of zoospores, while ammonia generated from decaying fish waste erodes the protective mucus on the skin (Chauhan *et al.*, 2014).

In freshwater fish, gills are essential organs for respiration and osmoregulation. Gills exhibit superior development in teleostean fishes compared to lungs in terms of efficiency. The gills of various fish exhibit significant modifications in response to alterations in oxygen environments and their functions, in conjunction with other respiratory structures for gas exchange. The indiscriminate application of insecticides in agriculture across diverse crops contaminates nearby water resources, including ponds and rivers, and severely harms aquatic fauna, particularly fish. The ingestion of pollutants adversely impacts fish, as their gills are particularly vulnerable due to their position and continuous contact with water, rendering them susceptible to harm from any irritating substances, whether dissolved or suspended (Lemke and Mount, 1963).

In the contamination process, heavy metals traverse biological barriers, including the gill epithelium and skin *via* a direct route, and the wall of the digestive system through an indirect route (Lloyd, 1992), accumulating mostly in metabolically active tissues such as the kidney, liver, and gills. Fish frequently occupy the apex of the aquatic food chain and can accumulate significant quantities of certain metals, which concentrate in their organs and disrupt essential physiological processes (Gomaa *et al.*, 1995). Chromium compounds are commonly found as environmental contaminants and are known to generate hazardous and mutagenic chemicals and their carcinogenic consequences (Kuykendall *et al.* 2006). Prabakaran *et al.* (2007) indicate that chromium (Cr) is an essential nutrient in biological systems. Chromium is regarded as both a heavy metal and a contaminant, in addition to being an important micronutrient. Chromium pollution in wastewater, resulting from electroplating, dyeing, tannery operations, and the production of hard-alloy and stainless steel, has adversely impacted life on Earth. Chromium serves as both a catalyst and a coating material (Idachaba *et al.*, 2004). The primary methods of introducing chromium into terrestrial environments include welding, grinding, and polishing of stainless steel, while its introduction into atmospheric and aquatic environments occurs through fossil fuel combustion and waste incineration (WHO, 1988). This pollution may impact all ecosystems and human health directly or *via* the food chain (Yılmaz *et al.*, 2009). Fish frequently occupy the apex of the aquatic food chain and can accumulate significant quantities of specific metals from water (Mansour and Sidky 2002). Besides adsorption on tissue and membrane surfaces, fish can assimilate metals through the intake of particulate matter or food in water, or *via* ion-exchange of dissolved metals through lipophilic membranes and gills (Mendil *et al.*, 2005). The distribution of metals throughout various tissues is contingent upon the route of exposure, specifically dietary and/or aqueous, and may act as an indicator of pollution (Alam *et al.*, 2002). This communication delineates the structural organization of the gill filament and gill lamellae epithelium of Puntius sophore, as well as the effects of chromium on several cellular components of these structures, employing a range of histochemical techniques (Singh *et al.*, 2010). These heavy metals are water-soluble and are absorbed by the body upon ingestion. Heavy metals can bind with enzymes, inhibiting their action and potentially causing serious physiological and neurological diseases. Heavy metals exhibit toxic, inhibiting, and mutagenic effects (Gourdon *et al.*, 1990). Heavy metals are defined as metallic elements with relatively high density that are fatal or hazardous at low quantities (Rana *et al.*, 2015).

The gills serve as the primary respiratory organs in fish, facilitating the exchange of oxygen and carbon dioxide through diffusion between blood and water. Gill slits, or gill clefts, are created by perforations in the lateral wall of the pharynx. The anterior and posterior walls of each gill cleft are elevated into vascular filamentous outgrowths, forming the gills, where the exchange of dissolved oxygen and carbon dioxide occurs. In addition to the gills, other accessory organs also function as respiratory structures in *C. Striatus* and *C. punctatus*. The histological structure of the gills in Gachua shows that the gills are externally covered by an operculum. They are of the holobronch type, characterized by a diminished interbranchial septum reinforced by cartilage. The gills consist of four pairs and two rows of primary gill lamellae, which are supported by the ceratobranchial and epibranchial segments of each gill arch. The interbranchial septum between the two rows of lamellae is short, allowing the distal ends of the lamellae to remain free.

The gill lamellae are reinforced by partially bony gill rays, which are linked to the gill arch and to one another by fibrous ligaments. Each gill ray is bifurcated at its proximal end, facilitating a passage for the efferent branchial vessel. Each gill arch accommodates two hemibranchiae or half gills. Each hemibranch consists of a series of elongated, slender filaments known as primary gill lamellae. Each primary lamella possesses several subsidiary lamellae on either side. The secondary lamellae are independent from one another but are united at the distal extremities of the major lamellae. Each secondary lamella comprises a central vascular layer encased in a thin layer of connective tissue and epithelium. The vascular layer contains a network of capillaries supported by pilaster cells. The flat, leaf-like structures serve as the primary site for gaseous exchange. The gill head is enveloped by a thin epithelium abundant in mucous glands. Consequently, the gill structure in fish facilitates close contact between water and the secondary lamellae for efficient gas exchange. In air-breathing species such as *C. Striatus* and *C. punctatus*, the pharynx forms a sac-like diverticulum to temporarily store air for gas exchange. The epibranchium of the first gill arch is enveloped by a delicate vascular respiratory membrane. It possesses a convoluted and intricately structured respiratory labyrinthine apparatus for supplementary respiration.

The fish liver resembles the liver of other vertebrates. It is a vital organ that regulates numerous life functions and significantly influences fish physiology, encompassing both anabolism (proteins, lipids, and carbohydrates) and catabolism (nitrogen, glycogenolysis, and detoxification), while also serving as a storage center for various substances, primarily glycogen (Takashima and Hibiya 1995; Ostaszewska and Sysa, 2004).

The liver is a digestive gland comprised of parenchymal cells and connective tissue fibers. It serves a significant function in metabolism and functions as a storage hub for many chemicals. Fish liver with a superior phylogenetic status exhibited structures analogous to the mammalian arrangement, which demonstrated enhanced metabolic activities. The liver of the fish is a comparatively substantial organ. In wild fish, the coloration is typically reddish-brown in carnivores and lighter brown in herbivores; nevertheless, it may appear yellow or off-white during specific seasons. Farmed fish may exhibit a lighter coloration compared to their wild counterparts, contingent upon dietary factors. The liver is a specialized organ in the front of the abdomen. In certain species, it may elongate the abdomen or closely adhere to other viscera. In certain animals, it is a composite organ known as a hepatopancreas. The fish liver contains drug-metabolizing enzymes and is among the most commonly impaired organs; yet, research indicates that only 10% of hepatic parenchyma is necessary to sustain adequate liver function in mammals. Fatty transformation (fatty metamorphosis; lipidosis) refers to an abnormal and excessive accumulation of intracellular lipids. An injury that, while reversible, can be significant and lead to considerable disruption of cellular function. Prominent non-staining vacuoles of adipose tissue reside in the cytoplasm, displacing and squeezing the nucleus. It is predominantly observed in the liver, but is also present in the kidneys and heart.

Bacterial and fungal infections induce metabolic crises and decrease liver function in fish. The reduction in protein, lipid, and glycogen levels indicates the exhaustion of cellular energy to satisfy the elevated demands of fish under stressful conditions. Consequently, the current investigation employed histochemical assays to localize the chemical products of cellular activity based on the degree of staining reactions. This is utilized to compare the protein, glycogen, and lipid levels in the liver cells of control and infected fish, *Channa striatus* and *Channa punctatus*.

Preliminary research was conducted in the domain of histochemistry on the livers of fish. The pancreas has been studied extensively in numerous fish species (Macallum 1884; 1886, and Kristal 1946). Ogura (1959) and Miyawaki *et al.* (1961) conducted histological and histochemical studies on the digestive gland of the crayfish, Procambarus clarkii. Subhash (1966) investigated the histochemistry of the liver and pancreas in the carp fingerling Catla catla. Raymond (1969) investigated the histology and histochemistry of the capsule of Proteocephalus sp. (Cestoda) in Lepomis macrochirus. David *et al.* (1972) examined the morphology and enzyme histochemistry of the liver in largemouth bass (*Micropterus salmoides*). Nakazawa *et al.* (1985) examined the histochemistry of liver tumors caused by diethylnitrosamine and the differential sex susceptibility to carcinogenesis in Oryzias latipes. Hampton *et al.* (1985) investigated the functional units in the liver of rainbow trout (Salmo gairdneri) concerning the layout and histochemical characteristics of hepatocytes. Krapagaganapthy *et al.* (1988) conducted histochemical experiments on the effects of lindane in the anabantid fish, *Colisa lalia*. Ribelles *et al.* (1995c) examined the morphological and histochemical alterations in the liver and pancreas of giltheads, Sparus auratus, induced by the acute effects of the anionic detergent sodium dodecyl sulfate, while Lynn *et al.* (1993) observed the biochemical and histochemical characteristics of hepatic tumors in rainbow trout, Oncorhynchus mykiss.

Rosety *et al.* (2001) investigated morphohistochemical alterations in the liver and intestine of juvenile gilthead, *Sparus aurata*, triggered by the acute effects of anionic tension active alkylbenzene sulphonate. Sakr *et al.* (2001) conducted histological, histochemical, and physiological investigations on the impact of the pesticide hostathion on the liver of the catfish *Glarias gariepinus*. Sakr and Jamal (2005) observed fenvalerate-induced histological and histochemical alterations in the liver of the catfish *Clarias gariepinus*, while Anandhi and Murthy (2006) investigated the chemical stress on the liver of *Glossogobius giuris*. Rejane and Vildes (2008) conducted ultrastructural and histochemical analyses of the liver of channel catfish (*Ictalurus punctatus*) subjected to fumonisin treatment. Pathan *et al.* (2009) delineated the histochemical alterations in the liver of the freshwater fish Rasbora daniconius, subjected to paper mill effluent. Presented the comparative histological, histochemical, and ultrastructural analyses of the liver in flathead grey mullet (*Mugil cephalus*) and sea bream (*Sparus aurata*) (Naveen and Hekmat, 2010).

The pancreas is primarily located between the two lobes of the liver. The pancreas in *C. Striatus* and *C. Punctatus* is thicker, pyramidal, and situated in the dorso-lateral regions. The pancreas has significantly large islets, which are round or oval structures encased in a thin layer of connective tissue. Numerous tiny blood capillaries are found within the connective tissue trabeculae of all the islets (Khanna and Singh, 1963).

The pancreas, a composite gland comprising exocrine and endocrine tissues, is found throughout the vertebrate lineage, exhibiting diverse morphological, anatomical, and cytological differences (Chakrabarti *et al.*, 2012). In teleosts, the pancreas, although less complex than that of higher vertebrates, performs many tasks, mostly facilitating digestion, while its endocrine component predominantly regulates glucose metabolism, according to Lazarow (1963). The pancreas in fish displays significant interspecies variation concerning its morpho-anatomical and cytological characteristics, as well as its distribution patterns, according to Weinreb and Bilstad (1955); Gammon *et al.* (1972); Sinha and Moitra (1975); Bremer (1980); Dwivedi and Pandey (1982); Ghosh (1996); Vicentini *et al.* (2005); Petcoff *et al.* (2006). Additionally, the variability in the cytological characteristics of the pancreas among fish with differing feeding behaviors has been examined by several researchers. Shafi (1973) noted a significant accumulation of zymogen granules in the acinar cells of the exocrine pancreas of *Ophiocephalus punctatus*, linking this trait to the fish's piscivorous diet. Examined the exocrine pancreas at various developmental stages of hybrid sturgeons and sought to clarify the correlation between eating behavior and the secretory function of the exocrine pancreas (Bremer, 1980).

The principal digestive gland in teleosts is the pancreas, which comprises exocrine and endocrine tissues exhibiting varying degrees of morphoanatomical and cellular alterations (Eurell and Haensly 1982; Youson *et al.*, 2006; El-said *et al.*, 2010). The presence of a hepatopancreas has been documented in the iridescent shark catfish, *Pangasius hypophthalmus*. Sayrafi *et al.* (2009) and a spleenopancreas in *Barbus pectoralis* (Khaksar Mahabaddy *et al.*, 2012).

The pancreas in fish may be extrapancreatic. In certain fish species, both characteristics are present in a single animal, as found in *C. Batrachus*, which possesses a pancreas comprised of two pairs of distinct lobes; the lateral lobes mostly consist of endocrine elements, while the median lobes form the hepatopancreas, as observed in *C. punctatus*. Benarjee (1986), Weber (1827), and Leguis (1873) were the first to document the existence of the pancreas in teleosts. The examination of the pancreas in fish has garnered relatively minimal focus due to the gland's widespread characteristics. Chesely (1934) examined the structure and function of the pancreas in various teleosts. He has documented the levels of protease, amylase, and lipase in specific marine fish species. Others who conducted investigations in teleosts include Smallwood and Derickson (1934), who studied *Cyprinus carpio*, whereas Byliss (1935) examined Phaceronectes. Sarbahi (1940) reported similar findings in *Labeo rohita*, Donaldson (1942) in Chinnook salmon, and Arguad and Gan (1945) in Anguilla. Sarbahi (1951) identified a hepatopancreas in *Carassius auratus*, a compact pancreas in *Micropterus salmoides*, and a widespread pancreas in *Ophiocephalus striatus*. Muthukkaruppan (1960) also reported similar observations.

The pancreas is a lobulated complex acinar gland, partitioned into two compartments with exocrine and endocrine activities, according to Mokhtar (2015). The pancreas is crucial for regulating energy balance and nutrition by synthesizing and releasing protein-digesting enzymes and hormones (Geyer *et al.*, 1996). The pancreas of teleost fish was categorized. Rizkalla (1967) categorized them as diffused, compact, and intrahepatic kinds. The dispersed variety is found surrounding the gallbladder and in the mesentery, whereas the compact pancreas is situated in the splenic mesentery, functioning as a distinct organ; the intrahepatic pancreas is placed within the liver. The exocrine pancreatic tissues associated with the liver near the portal vessels are collectively referred to as hepatopancreas (Ferguson, 1989). Throughout ontogenesis, the exocrine pancreatic tissue forms around the portal vein and subsequently infiltrates the liver parenchyma or remains extra-hepatic, contingent upon the species (Gonzalez *et al.*, 1993). The grass carp, also known as white amur (*Ctenopharyngodon idella*), is a sizable cyprinid fish indigenous to major Asian rivers, including the Amur River Basin in Russia and the West River in China, according to Gulliary & Gasaway (1978). It is a rapidly proliferating herbivorous fish that mostly consumes grass or other water flora, making it an advantageous species for aquaculture in Egypt (Earl *et al.*, 1992).

In freshwater fish, the pancreas is primarily diffuse, with exocrine pancreatic cells located around the major capillaries. The endocrine tissue is structured as Langerhans islets, whereas the exocrine serous cells are arranged in acini, constituting the parenchyma dispersed throughout the omentum. A thin layer of dense pancreatic tissue, not readily visible macroscopically, is situated along the ventral surface of the stomach, adjacent to the extra-hepatic bile duct and several major arteries, extending to the loop of the midgut.

Pancreatic tissue mostly consists of exocrine acini, which are formed of pyramidal-shaped acinar cells. The endocrine component of the pancreas is sparsely interspersed with exocrine acini. The islets of Langerhans are rather big and consist of three cell types, with beta (β) cells thought to secrete insulin. These β cells are polyhedral in shape, featuring a large, spherical central nucleus.

The two other types of cells are alpha (α1) and alpha s (α2). In fish, α1 cells are more abundant than α2 cells. The α1 cells are also referred to as argyrophils, whereas α2 cells are classified as non-argyrophils. These cells are approximately triangular in shape with an eccentric nucleus; one of the two alpha cells is thought to release glucagon.

The objective of the current investigation was to delineate the *C. Striatus* and *C. Punctatus* pancreas examined under light microscopy, emphasizing histochemistry and enzyme activity of pancreatic acini, with particular focus on the location of islets of Langerhans.

Materials and Methods

The freshwater fish specimens for this study were procured from various lakes in Hasanparthy, Dharmasagar, and Bandham in the Warangal district of Telangana, India. The EUS-infected fish samples of *Channa striatus* and *Channa punctatus* were collected and maintained as previously described in the first chapter. The specimens were captured alive using a fishing net and promptly transported to the laboratory in plastic containers filled with oxygenated water. The specimens were profoundly anesthetized by immersion in a 5 ml/L water solution of ethylene glycol monophenyl ether. To examine the gross anatomy of several locations, tissue samples from the skin, gills, liver, and pancreas were meticulously excised, and minute fragments were collected for optimal fixation. For the histopathological and histochemical analyses, the necessary tissues were excised, and the associated blood vessels and mucus were meticulously scraped away without compromising the original structure. The tissues obtained for this investigation comprised skin, gills, liver, and pancreas. The methodology employed for the production of histological sections and for histochemical examination of various substances is consistent across all tissues. The tissues obtained from both control and diseased fish were promptly fixed and processed. The fixatives employed in this investigation were Alcoholic Bouin's, Susa, Formal Calcium, Carnoy, and Zenker. Particular fixatives were employed for lipids, mucopolysaccharides, and nucleic acids. Formalin was utilized for lipids, while Susa and Carnoy were employed for nucleic acids. Modified Bouin's solution, as described by Harris *et al*. (1973), and a 1% cetylpyridinium chloride solution in formalin were specifically used for mucopolysaccharides. Bouin's and Susa had favorable outcomes in comparison. Following fixation in Bouin's solution for 18-24 hours, the sample was rinsed in Scots tap water. The Susa fixed substance has undergone treatment with iodine alcohol. Calcium-fixed formalin material exposed to post-chromatin treatment has been utilized for lipid analysis in tissue sections. Experience demonstrated that Susa served as the optimal fixative for the majority of histological and histochemical reactions, excluding lipids, nucleic acids, and mucopolysaccharides. Subsequent to fixation and post-treatment, the sample was washed and subsequently dehydrated using a graded sequence of alcohols. Dehydrated with xylene and encapsulated in paraffin wax. The infiltration time varied from 1 to 4 hours, depending on the system components, after which serial transverse or sagittal slices were cut at a thickness of 3-6 μm for all tissues. Optimal staining outcomes for histological analyses can be achieved with the application of Dalafield's haematoxylin counterstained with eosin (Gurr, 1962) and Heidenhain's Azan (Gurr, 1962).

The EUS infection investigations involved fish gathered from three lakes: Hasanparthy, Dharmasagar, and Bhandham, comprising two species: *Channa striatus* and *Channa punctatus*. The protocols for identifying infected and control fish involved the separate collection and individual processing of tissues from both groups for histopathological and histochemical analyses. Various histochemical approaches have been utilized to elucidate the chemical composition, specifically the presence of carbohydrates, proteins, lipids, phospholipids, and nucleic acids in diverse tissues such as the skin, gills, liver, and pancreas of *Channa striatus* and *Channa punctatus*. The methodology described by Pearse (1968) for various histochemical assays was utilized in the current investigation. Nevertheless, the methodologies delineated in the subsequent texts were also cited (Gomori 1952; McManus and Mowry 1960; Lillie 1965; Humason 1967; Bancroft 1975; Carleton and Drury 1957; Barka and Anderson 1963; Chayen *et al.* 1973; Culling 1974; Bancroft and Steyans 1977; Kiernan 1999; Shyamasundari and Hanumantha Rao 2007).

Carbohydrate Identification

Carbohydrates were tested using material fixed in Bouin's and Susa fixatives. The Periodic Acid/Schiff (PAS) approach (Pearse 1968) is utilized to identify carbohydrates and other groups. Periodic acid induces oxidative cleavage of carbon-carbon bonds in 1, 2 glycols, resulting in the formation of dialdehydes, which are then stained by Schiff's reagent. This oxidant does not further oxidize the resultant aldehydes. Given that several compounds yield a positive PAS reaction, the distinct PAS-positive groups present were further defined using various techniques applied to the sections.

Polysaccharides

The traditional technique for identifying polysaccharides uses the Periodic Acid Schiff (PAS) approach. Given that various compounds are known to yield a positive reaction using the PAS approach, appropriate controls were utilized to identify the specific component responsible for the positive reaction. The procedure was conducted without preceding oxidation using periodic acid to ascertain if the response was attributable to pre-existing aldehydes for glycogen detection; sections underwent the PAS light green technique alongside diastase digestion. Best's caramine technique was employed to test the existence of glycogen. The PAS reaction was performed following acetylation (24 hours at ambient temperature in 16 ml of acetic acid and 24 ml of pyridine) and subsequent deacetylation (45 minutes in 0.1 N potassium hydroxide at ambient temperature or with 20% ammonia in 70% alcohol for 24 hours) to confirm the presence of 1:2 glycol groups. To ascertain if the reactivity is attributable to lipid, sections were subjected to treatment with various lipid solvents, including pyridine or a methyl alcohol-chloroform mixture, prior to the application of the

PAS technique. If the substance exhibited a positive PAS reaction that was not extractable by diastase or lipid solvents, and also yielded a positive result in protein tests, it was classified as either a mucoprotein or glycoprotein. The identity was ultimately verified using the methylene blue extinction procedure, which entails staining portions with methylene blue at several pH levels.

Mucous Substances

Tissue was fixed in either new Comer's fluid dioxane or 1% acetyl pyridinium chloride in 10% formaldehyde containing 2% calcium acetate, as modified by Bouin's Harris *et al.* (1973). Among all these acetylpyridinium chlorides, formalin was the most effective, as determined by cell integrity and the staining intensity, particularly for mucopolysaccharides. A range of histochemical assays was utilized to identify various mucous cell types and to define the mucosubstances they produce, based on their vicinal hydroxyl groups, carboxyl groups, sulfate groups, or a combination thereof.

To distinguish mucosubstances, sections underwent the following tests: the PAS technique as described by McManus (1946), the PAS technique with prior acetylation as per McManus and Cason (1950), the acetylation-deacetylation method by McManus and Cason (1950), PAS following diastase digestion, and PAS subsequent to phenylhydrazine treatment (5% phenylhydrazine for 1 hour at 25°C) (Spicer *et al.*, 1967). These serve to illustrate mucosubstances containing vicinal hydroxyl groups.

Acid mucosubstances were identified utilizing the following methodologies: Alcian blue (AB) at pH 2.5 (1% AB 8GX in 3% acetic acid) for 30 minutes, AB at pH 1.0 (1% AB in 0.1N HCl) for 30 minutes, and Mowry's (1956); Lev and Spicer's (1964); and Mowry's (1963) adaptations of Male's colloidal iron solution for 2 hours.

Acid mucosubstances were differentiated from neutral mucosubstances using specific tests, utilizing a combination technique of PAS and AB at pH 2.5 (Mowry and Winkler, 1956) and at pH 1.0 (Spicer *et al.*, 1967).

The subsequent processes were implemented to differentiate sialomucins from sulfomucins. Aldehyde Fuchsine (AF) as described by Halmi and Davis (1953), AF / AB (pH 2.5) according to Spicer and Mayer (1960).

The majority of the aforementioned approaches were executed concurrently with additional procedures and specific assays using chemical inhibition or enzymatic elimination of particular reactive groups in the mucosubstances. The procedures encompass mild methylation (0.40 cc of concentrated HCl in 50 cc of 100%

alcohol for 4 hours at 37°C), succeeded by Alcian blue staining at pH 2.5, and active methylation (4 hours at 60°C) followed by Alcian blue staining at pH 2.5 (Fisher and Lillie 1954; Spicer 1960). Mild methylation saponification (1% KOH in 70% alcohol for 20 minutes) was conducted, followed by AB staining (pH 2.5) and active methylation saponification as per Spicer and Lillie (1959). These methylation and demethylation procedures were executed alongside Azure A, AB/PAS, and AF/AB treatments.

PROTEINS

Mercuric Bromophenol Blue

The existence of basic proteins in the tissues was confirmed through the application of mercuric bromophenol blue. The methodology employed was that of Mazia *et al.* (1953). Bromophenol blue is an anionic dye that associates with protein cations in the presence of salts such as mercuric chloride. It interacts with both anionic and cationic protein groups, resulting in a blue coloration. This assay was concurrently validated by control slides using Vanslike's reagents.

Further experiments have been undertaken to demonstrate the precise type of proteins.

Ferric Ferricyanide Method (FFC)

This approach by Chevermont and Frederic (1943) was employed to identify sulfhydryl groups. This approach is based on the premise that sulfhydryl groups act as potent reducing agents, converting ferricyanide to ferrocyanide. The ferrocyanide subsequently interacts with ferric ions to yield a vivid and insoluble precipitate of Persian blue.

Millon's Response

This approach demonstrates the phenolic amino acid, tyrosine. This is an advantageous marker for the majority of proteins. The prevalent histochemical method for detecting this amino acid has been an adaptation of Millon's test, a protocol specifically devised and refined by Baker (1956). In this process, phenol is transformed into nitrosophenol by sodium nitrate, and mercury binds to the nitrogen atom of the nitrous group, resulting in a red-colored molecule.

The other assays employed for the identification of specific proteins in tissues are as follows

P-Dimethyl amino benzaldehyde nitrate (P-DMAB Nitrate) as per Adam's methods (1957) for tryptophan.

Performic acid/alcian blue method of Adams and Sloper (1956) for detecting disulfide groups.

$KMnO_4$/Aalcian blue as described by Arvy and Gabe (1962) for the detection of disulfides.

Ferric ferricyanide, as described by Cheveremont and Frederic (1943), for sulfhydryl groups.

Ninhydrin/Schiff method by Yasuma and Itchikawa (1953) for detecting protein-bound NH2 groups.

Congo red staining method after Pearse (1968) for the identification of glycoproteins.

Primarily, material preserved in Susa, Zenker's solution, or alcoholic Bouin's fixative was utilized. The techniques employed for proteins include the xanthoproteic method, the Bromophenol blue technique by Mazia *et al.* (1953), Azure/Schiff by Kasten (1960), and ninhydrin/Schiff as described by Pearse (1968). Deamination was performed using Vanslike's reagent. Disulfide groups were seen by previous oxidation with performic acid or acidified potassium permanganate, followed by staining with AB, Schiff's reagent, or Scott (1953); Adams and Sloper (1955); Pearse (1961). Sulfhydryl groups were identified using Dihydroxy Dinaphthyl Disulfide (DDD) as per Barnett and Seligman (1954) and the ferric ferricyanide method described by Chevremont and Frederic (1943).

Proteins comprising elastin, collagen, and keratin were seen using the following methods: PAS, AF, Unna's Orcein, Verhoeff's, Safranin O, Spirit blue, and Elder and Owen 1967, to ascertain the existence of elastin. Collagen was identified by using Luxol fast blue G in methanol, as per Salthouse (1965), Gurr (1962), and Aniline blue methodologies. Keratin was localized using $KMnO_4$/AB, performic acid/AB, and Rhodamine B methods.

Characterization of Lipids

Optimal results may be achieved using material preserved in Baker's (1956) formal calcium and subsequently processed with post-chromatin for the identification of lipids and phospholipids. Histochemically, lipids are characterized as fatty substances exhibiting certain chemical properties of lipid molecules. There are two categories of testing for lipid demonstration: physical methods and chemical approaches. The physical approaches are typically more significant as lipids are predominantly identified histochemically by their capacity to concentrate Sudan Black B.

Sudan Black B Method

Sections for lipid staining were exposed to Sudan Black B for a duration of 30 to 60 minutes. The Sudan black B method, utilizing a saturated solution of Sudan black B in 70% alcohol, as described by Chiffellie and Putt (1951), was employed to identify the presence of lipids. Sudan Black is a fat-soluble dye that dissolves in pure alcohol but is insoluble in water. This exhibited a strong attraction for lipid substances, particularly phospholipids. The bright coloration and significant affinity of Sudan Black B for various lipid types render it a valuable histochemical criterion for identifying lipid content in tissue components. The black color signified the presence of lipids. Sudan Black B exhibits a significant affinity for both hydrophilic and hydrophobic lipids. The Kluver and Barrera (1953) approach, utilizing copper phthalocyanin, was also employed effectively to ascertain the existence of phospholipids. Controls for lipids were obtained by the pyridine extraction process and Kelling's extraction utilizing either cold or hot acetone or a methyl alcohol-chloroform mixture (Baker, 1946).

Identification of Nucleic Acids

To examine the distribution pattern of nucleic acids, tissues were fixed using Carnoy's fixative. The existence of DNA was evidenced by Feulgen's reaction (hydrolysis in 1 N HCl for 15 minutes at 60ºC) as described by Pearse (1968). The Methyl green/Pyronin Y approach developed by Pearse (1968) was employed to simultaneously detect DNA and RNA in the same section, whereas RNA was exclusively identified using Pyronin Y alone.

Methyl Green-Pyronin Y Technique

The application of a combination of the basic dyes methyl green and pyronin-Y for the differential staining of nuclei and cytoplasm is a longstanding technique. Nonetheless, the mechanism *via* which this mixture induces differentiation has been elucidated by Chayen (1973) and Barnett (1954). The mechanism of this reaction appears to include competition between the sluggish staining of doubly charged methyl green and the relatively quick staining of singly charged pyronin-Y. Consequently, methyl green preferentially stains DNA, while pyronin-Y stains RNA.

Feulgen's Reaction

The specific chemical process with this approach remains ambiguous. This may pertain to the capacity of the phosphate to translocate to various locations inside the sugar molecule. Mild hydrolysis (1 N HCl at 60°C for several minutes) reveals the aldehyde groups. This hydrolysis process eliminates purines from DNA, hence increasing the availability of aldehyde groups for staining. Nevertheless, the extended hydrolysis makes the DNA soluble. The aldehyde groups are subsequently treated with Schiff's reagent, resulting in a color change.

Microphotographs of several tissue sections have been captured at low and high magnification to elucidate the changes in both structural and histochemical representations.

OUTCOMES

Comparative Histochemical Examination of Control and Diseased Dermal Tissue

A number of histochemical approaches have been employed to elucidate the chemical composition of the chemicals in both control and infected skin of *C. Striatus* and *C. Punctuated.* The skin tissue comprises flat cells, goblet or club cells, mucin-producing cells, dense collagenous connective tissue or coarse collagen fibers, melanin-containing cells, and free red blood cells, all of which typically exhibit analogous responses to histochemical reactions. The staining intensity can be utilized to compare the protein, lipid, and glycogen levels in the cells of control fish with those of infected fish at varying concentrations. This study discusses the responses of different cells to the applied histochemical assays and their characteristics.

GLYCOGEN CONCENTRATION

Periodic Acid-Schiff (PAS) Reaction

In the control skin, all epidermal cells exhibited mild PAS staining. The goblet cells, also known as club cells, exhibit mild staining. The reaction in the flat cells is more powerful than in the other cells.

In the infected skin tissue, collagenous fibers had a robust positive reaction to the Periodic Schiff test, indicating the presence of glycogen compounds.

Alcian Blue (Ab) pH 2.5

This approach is utilized to identify the presence of acidic mucopolysaccharides. At pH 2.5, the control skin tissue exhibited a pale blue coloration in the presence of acid, while the goblet cells demonstrated a negative reaction. All mucous gland cells and flat cells exhibited a positive reaction, indicating the presence of weakly acidic mucopolysaccharides, hyaluronic acid, and sialomucins.

In diseased dermal tissue, several regions exhibited light and dark blue bands. The presence of sulphated mucins was established.

Alcian Blue (Ab) pH 1.0

At pH 1.0, the control skin tissue cells had a somewhat blue hue, signifying the presence of acidic sulfated mucosubstances. An adverse reaction was observed in the fiat cells and some goblet cells. The majority of the mucous glands exhibited a weakly positive reaction.

Infected skin tissue treated with AB at pH 1.0 exhibited greenish-blue staining in the superficial and outer middle cell layers. These results showed the existence of glycoproteins with carboxyl groups and/or sulfate esters in low amounts at these sites (Plate **I**, Figs. **1** to **12**).

PLATE - I

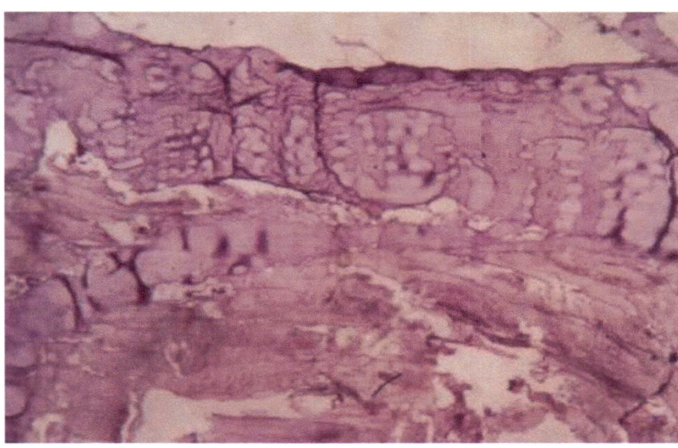

Fig. (1). T.S of *Channa striatus* control skin shows moderate presence of glycogen (PAS).

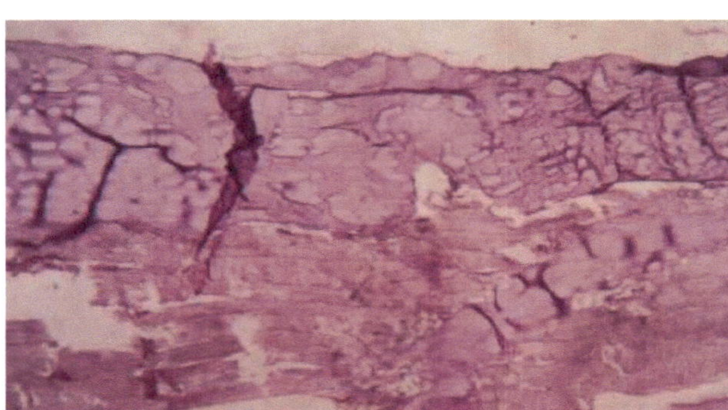

Fig. (2). T.S of *Channa striatus* infected skin shows a strong presence of glycogen substances (PAS).

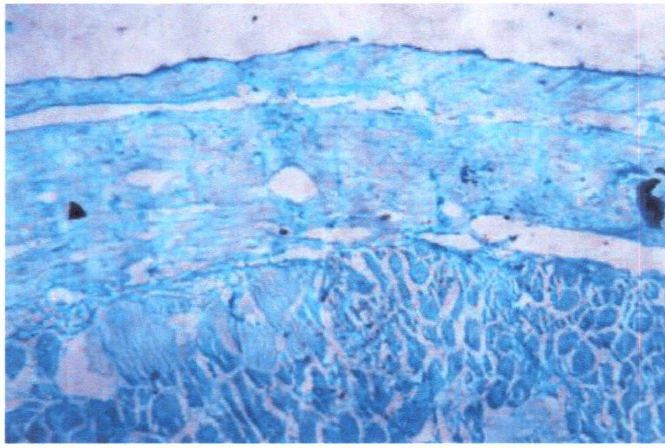

Fig. (3). T.S of *Channa striatus* control skin shows pale blue presence of acid mucin (AB2.5pH).

Fig. (4). T.S of *Channa striatus* infected skin shows light and dark blue presence of sulphated mucins (AB 2.5 pH).

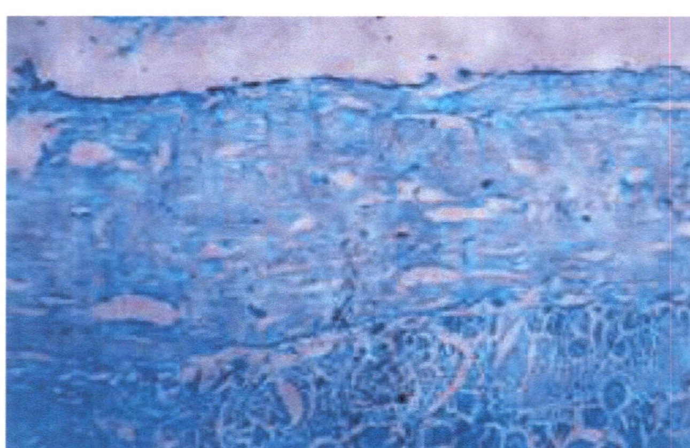

Fig. (5). T.S of *Channa striatus* control skin shows the presence of mucin (AB 1.0 pH).

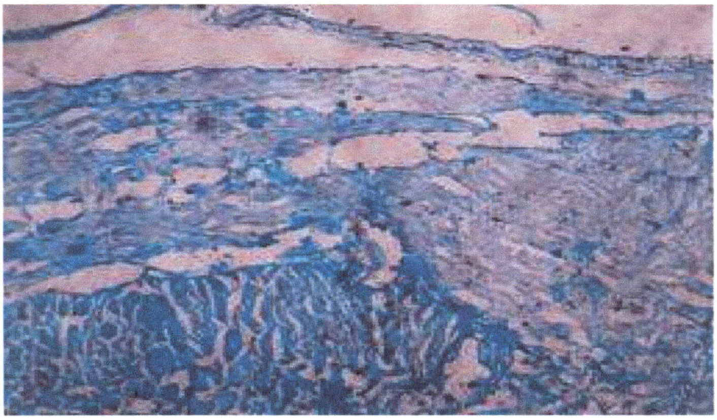

Fig. (6). T.S of *Channa striatus* infected skin shows the presence of sulfated mucosubstances (AB 1.0 pH).

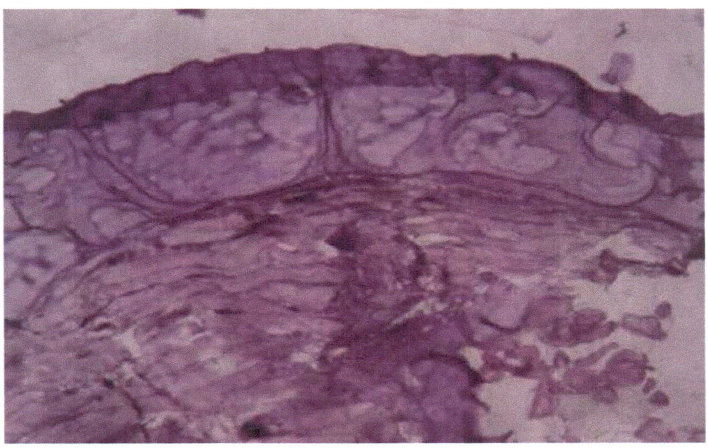

Fig. (7). T.S of *Channa punctatus* control skin shows moderate presence of glycogen (PAS).

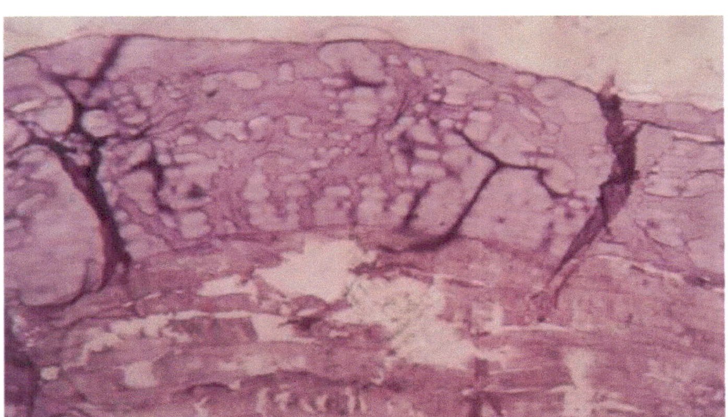

Fig. (8). T.S of *Channa punctatus* infected skin shows a strong presence of glycogen substances (PAS).

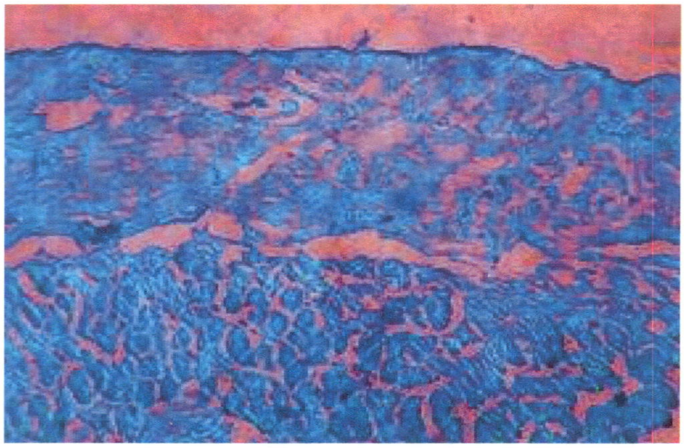

Fig. (9). T. S of *Channa punctatus* control skin shows a pale blue presence of acid mucin (AB 2.5pH).

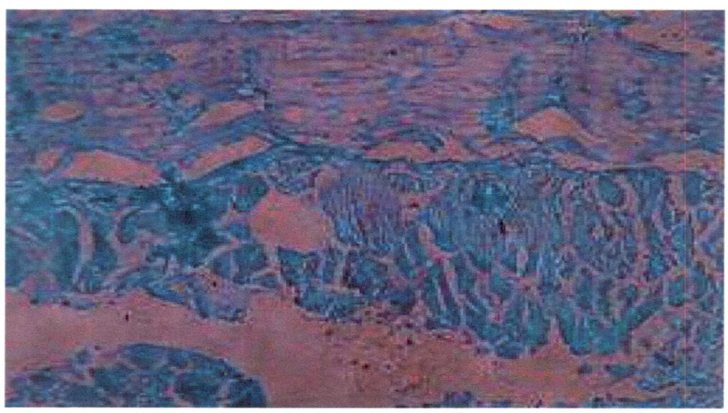

Fig. (10). T.S of *Channa punctatus* infected skin shows light and dark blue presence of sulphated mucins (AB 2.5 pH).

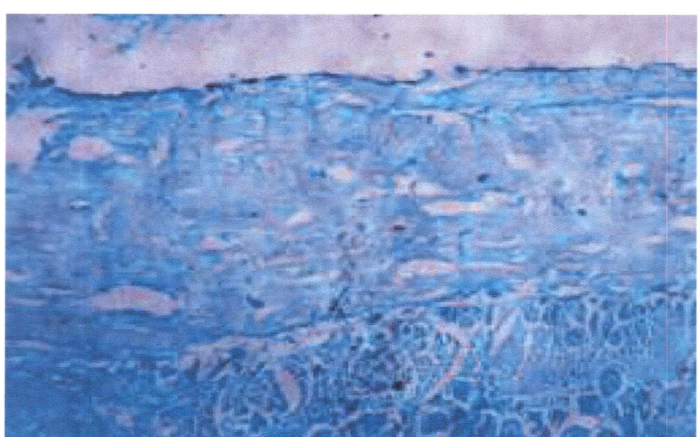

Fig. (11). T.S of *Channa punctatus* control skin shows the presence of mucin (AB 1.0 pH).

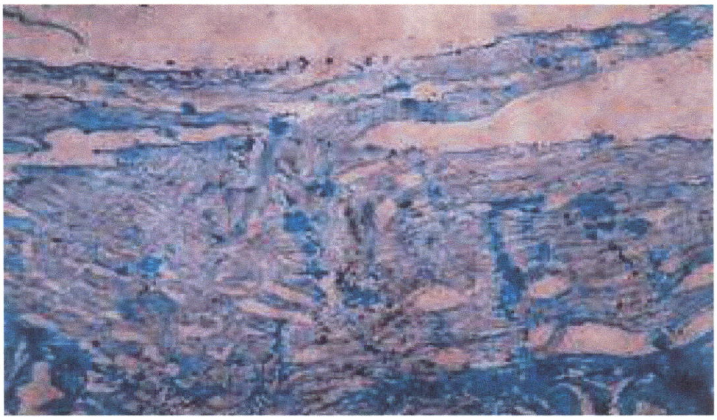

Fig. (12). T.S of *Channa punctatus* infected skin shows the presence of sulfated mucosubstances (AB 1.0 pH).

Alcian Blue pH 2.5 / PAS

The amalgamation of the AB (2.5pH) and PAS techniques reveals a compelling outcome. The mucosubstances were dyed red using the Alcian blue (pH 2.5)/PAS procedure. Neutral mucopolysaccharides that stain blue are hyaluronic acid and sialomucins, while those that exhibit a blue-purple stain comprise both neutral and weakly acidic mucopolysaccharides.

The control skin had a pink hue indicative of predominant glycogen content, along with a faint blue coloration suggesting a lesser quantity of mucin. The planar cells exhibited no coloration. The goblet cells exhibit a blue coloration, signifying the presence of hyaluronic acid and sialomucins. It is important to note that these cells have also exhibited a favorable response to PAS. Mucous gland cells appeared to harbor carbohydrate-protein complexes called mucoproteins. The dense collagen connective tissue had a blue-purple hue, indicating the concurrent presence of neutral mucins and mildly acidic mucopolysaccharides. Some melanin cells, however, exhibit a blue hue. As no protein was detectable in these cells, it is hypothesized that they contain solely mildly acidic mucopolysaccharides.

Infected skin exhibited an increase in blue, indicating the presence of mucin, with a corresponding rise in the blue component.

Aldehyde Fuchsine

The control skin exhibited purple elastic fibers and light yellow-stained collagen when treated with aldehyde fuchsine, while the flat cells displayed a negative reaction. Goblet cells exhibit a vivid purple hue. The red coloration of the mucous cells indicates the probable presence of sulfomucins.

The infected skin tissues had a robust positive reaction to Aldehyde fuchsine, indicating the presence of sulfated mucosubstances. Conversely, the diseased skin tissue exhibited a reduction in elastin levels (Plate **II**, Figs. **1** to **8**).

PLATE – II

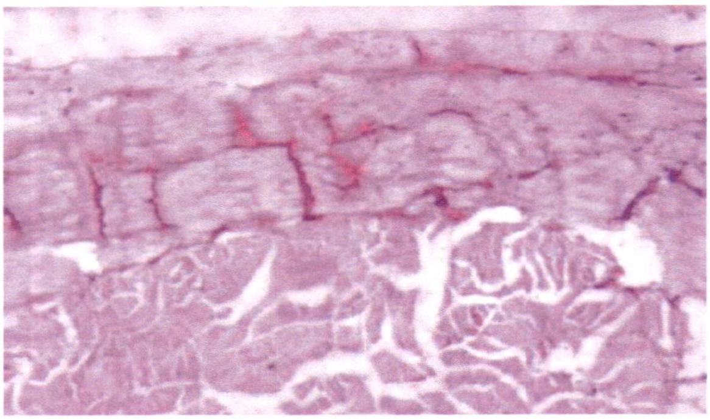

Fig. (1). T.S of *Channa striatus* control skin, displaying the presence of glycogen predominantly (pink), a lesser amount of mucin (blue) (AB 25pH/PAS).

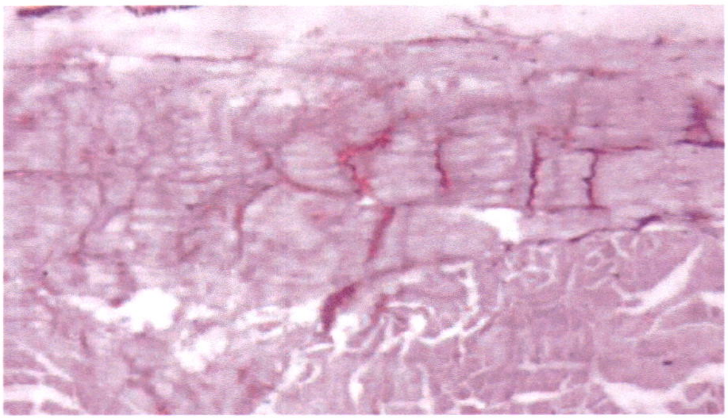

Fig. (2). TS of *Channa striatus* infected skin displays the presence of mucin predominantly (AB 2.5 pH /PAS).

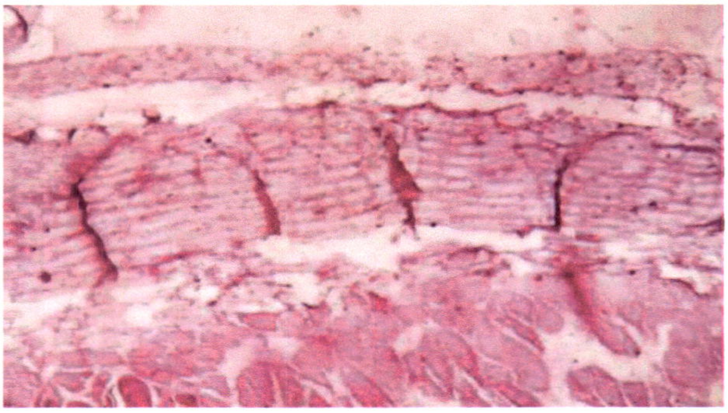

Fig. (3). T5 of *Channa striatus* control skin presents elastic fibers stained purple, and collagen stained with light yellow Aldehyde fuchsine.

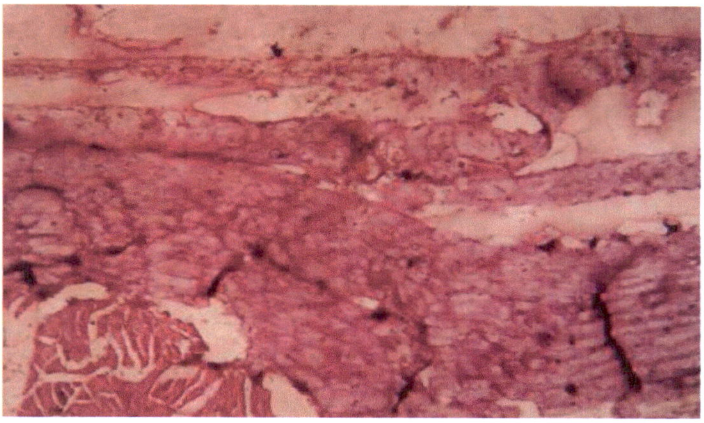

Fig. (4). TS of *Channa striatus* infected skin shows decreased Aldehyde fuchsine.

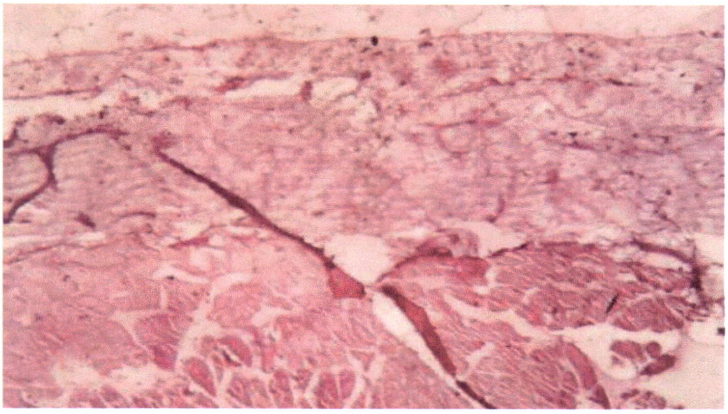

Fig. (5). T.S of *Channa punctatus* control skin, displays the presence of glycogen predominantly (pink), a lesser amount of mucin (blue) (AB 25pH/PAS).

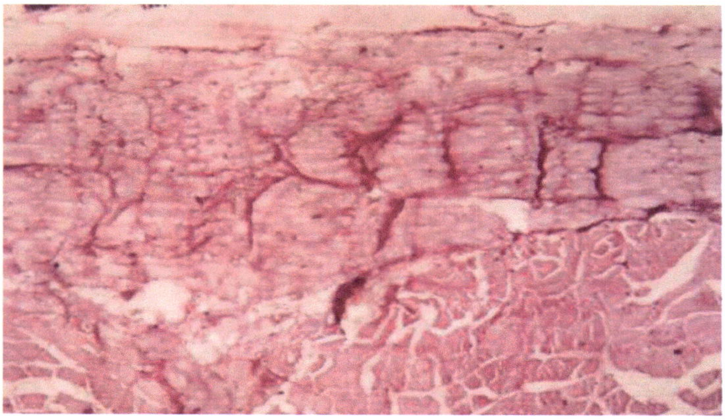

Fig. (6). T.S of *Channa punctatus* infected skin displays the presence of mucin predominantly AB 2.5 pH /PAS.

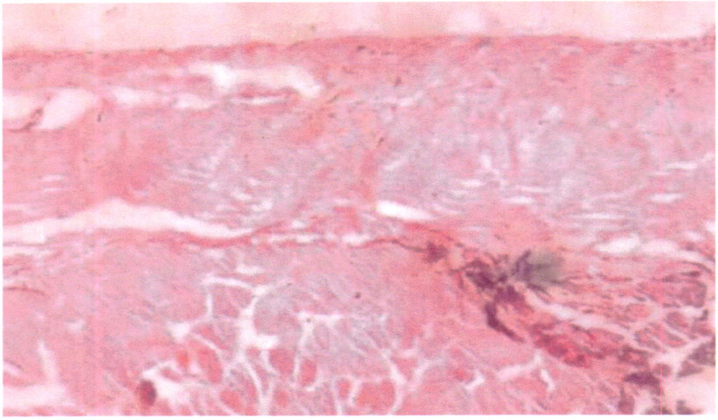

Fig. (7). T.S of *Channa punctatus* control skin presented elastic fibers stained purple, and collagen stained with light yellow (Aldehyde fuchsine).

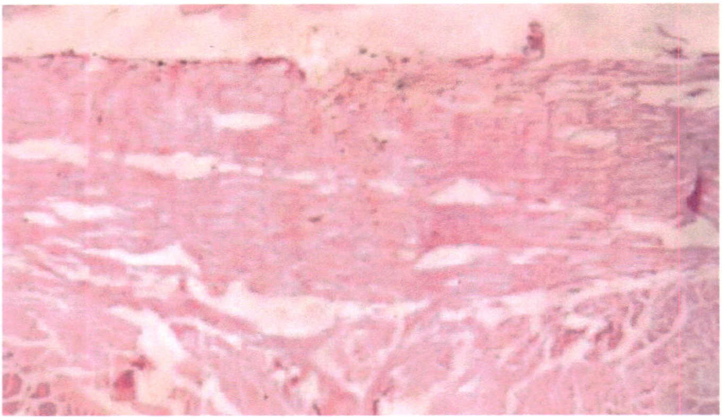

Fig. (8). T.S *Channa punctatus* infected skin shows decreased Aldehyde fuchsine.

PROTEINS

Mercuric Bromophenol Blue

For protein testing in tissue slices, Susa and Bouin's fixed materials have been utilized. A substantial quantity of testing has been conducted to examine the presence of various protein types.

This assay is utilized to identify the existence of fundamental proteins. Bromophenol blue indicated a positive reaction in flat cells and goblet cells, while bundles of coarse collagenous fibers stained blue, suggesting the presence of basic proteins.

The infected skin tissue, which is tinged green, indicates the depletion of fundamental proteins. Total protein showed a significant reduction in flat cells, goblet cells, and collagenous fibers.

The histochemical techniques employed in this work have unequivocally demonstrated that in the two kinds of fish. The skin includes a substantial quantity of proteinaceous substances. This was confirmed when the skin slices were submitted to additional specialized stains for the detection of various proteins (Plate III, Figs. **1** to **4**).

PLATE – III

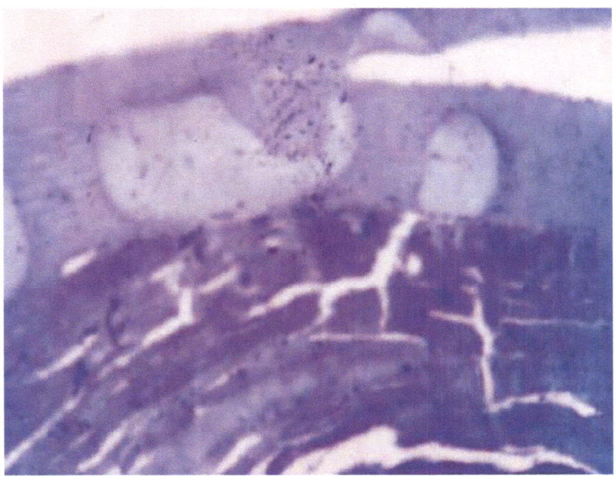

Fig. (1). T.S of *Channa striatus* control skin displaying the presence of basic proteins (Bromophenol blue).

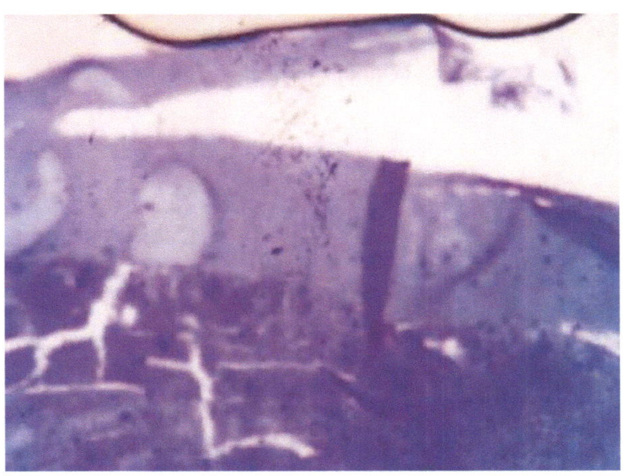

Fig. (2). T.S of *Channa striatus* infected skin displaying depletion of basic proteins (Bromophenol blue).

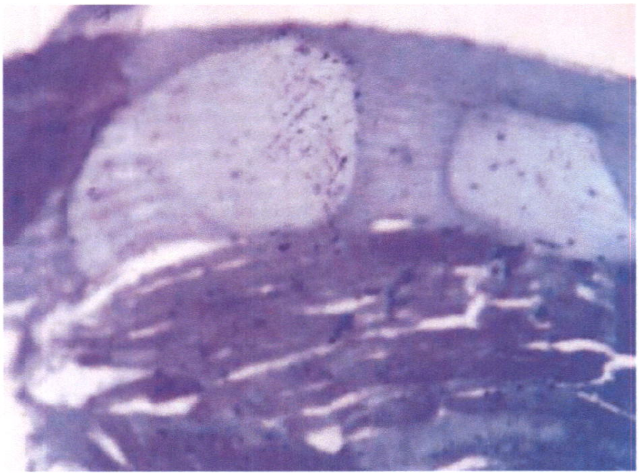

Fig. (3). T.S of *Channa punctatus* control skin displaying the presence of basic proteins (Bromophenol blue).

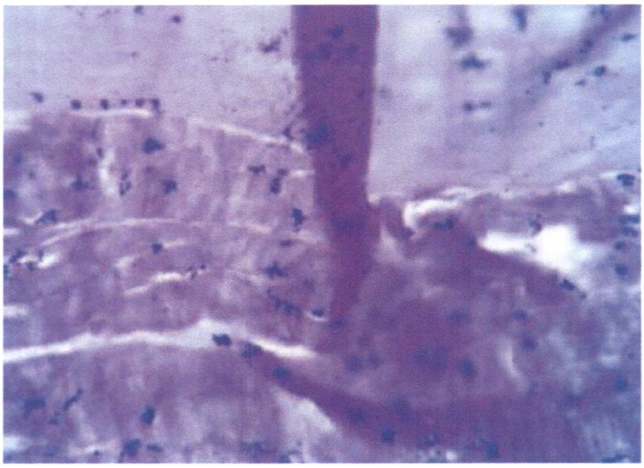

Fig. (4). T.S of *Channa striatus* infected skin shows light and dark blue presence of sulphated mucins (AB 2.5 pH).

Congo Red

This approach is utilized to illustrate the existence of glycoproteins. A pronounced positive was noted in the typical skin tissue. This positive indicated the existence of protein-bound amino groups and glycoproteins within the cells. These results show that glycoproteins are the principal components of epidermal cells. Only the squamous cells and club cells had a pronounced positive effect. This observation, supported by the responses of these cells to other previously utilized stains, indicates the involvement of glycoproteins.

Infected skin tissue and cells treated with Congo red exhibited a somewhat positive reaction. It was entirely devoid of protein, with remains primarily situated inside the collagenous fibers (Plate **IV**, Figs. **1** to **4**).

PLATE – IV

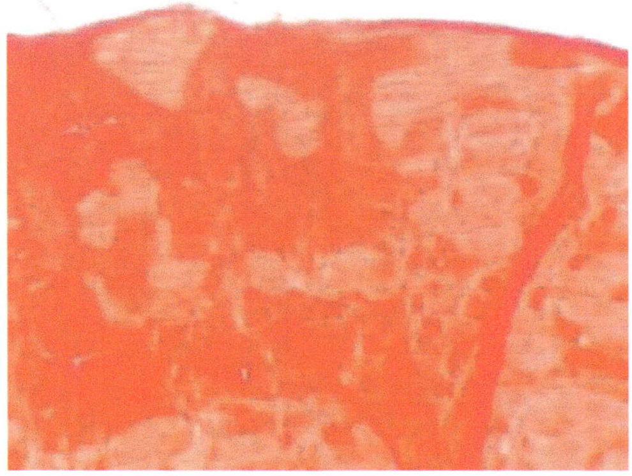

Fig. (1). T.S of *Channa striatus* control skin displaying the presence of protein-bound amino groups and glycoproteins in the cells (Congo red).

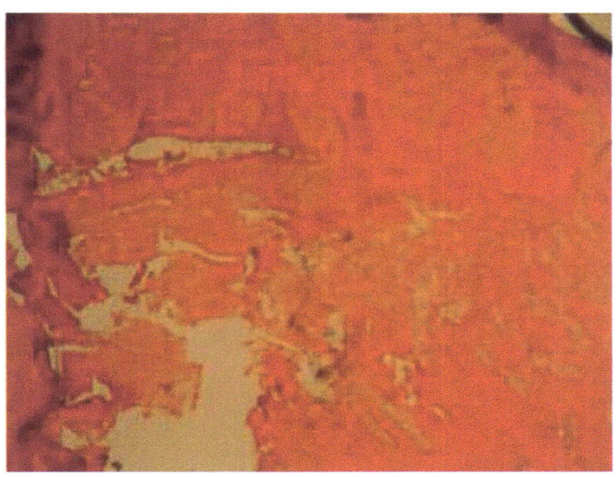

Fig. (2). T.S of *Channa striatus* infected skin displaying reduction of glycoproteins (Congo red).

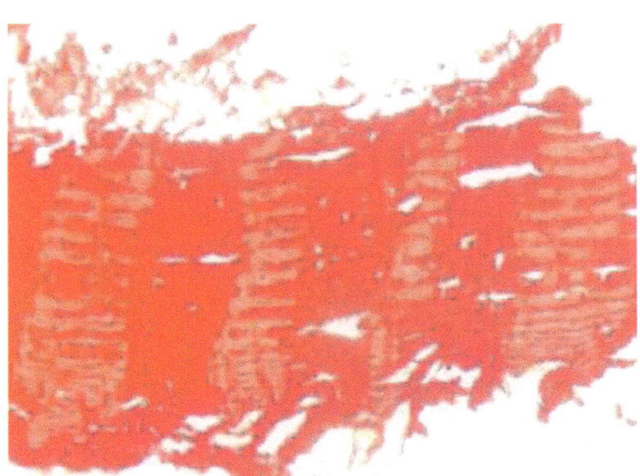

Fig. (3). T.S of *Channa punctatus* control skin displaying the presence of protein-bound amino groups and glycoproteins in the cells (Congo red).

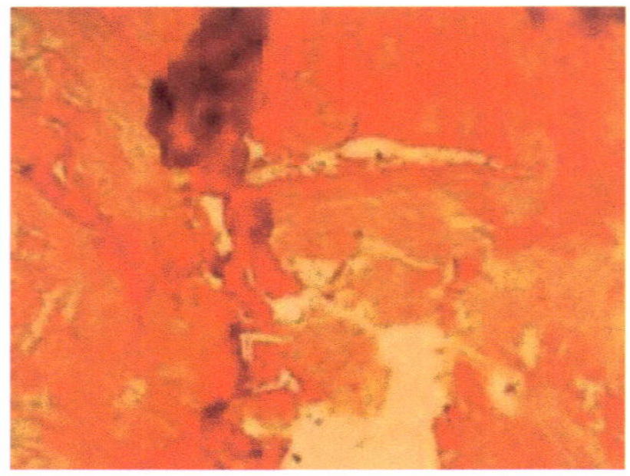

Fig. (4). T.S of *Channa punctatus* infected skin displaying reduction of glycoproteins (Congo red).

LIPID COMPOSITION

Sudan Black B

Calcium-fixed tissue was sectioned and stained with Sudan black B. The glands demonstrated significant positive results, indicating the presence of lipids. This staining was eliminated when sections underwent pyridine extraction, corroborating the existence of lipids in the glands. The glands tested positive for Copperphthalocyanin, indicating the presence of phospholipids. This approach allows for the classification of the glands as lipid glands.

In the infected skin tissues treated with Sudan black B, collagenous fibers exhibited a moderate positive reaction, indicating the presence of lipids and phospholipids.

The histochemical reactions revealed that the normal skin tissue of this fish comprised glycogen, weakly acidic sulfated mucosubstances, hyaluronic acid, sialomucins, carboxylated mucosubstances, basic and acidic proteins, sulfide and sulfhydryl groups of proteins, protein-bound amino groups, glycoproteins, lipids, phospholipids, elastin, and collagen (Plate **V**; Figs. **1** to **4**).

PLATE – V

Fig. (1). T.S of *Channa striatus* control skin tissue exhibiting the presence of lipids (Sudan black B).

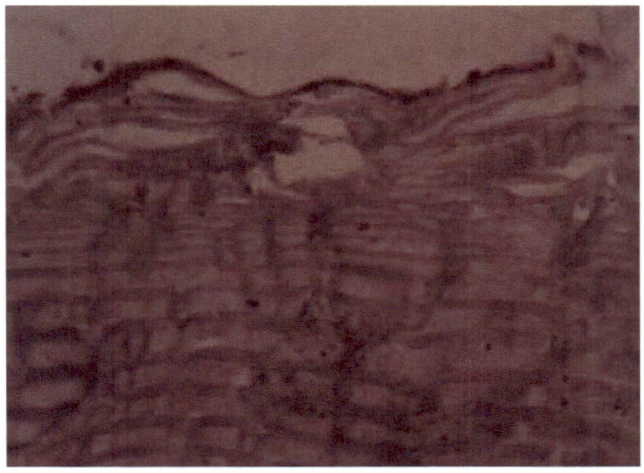

Fig. (2). T.S of *Channa striatus* infected skin tissue displaying the lipids and phospholipids (Sudan black B).

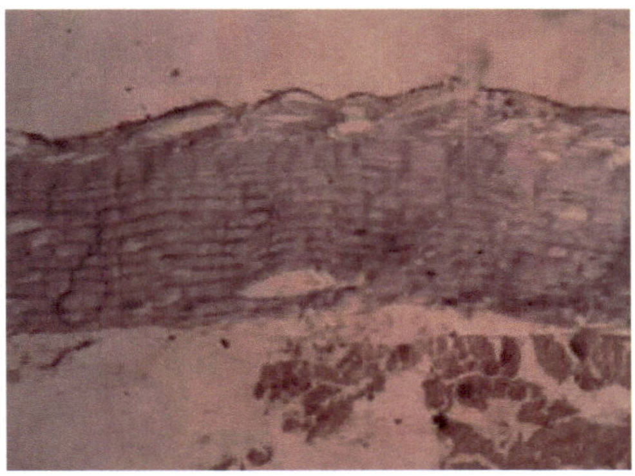

Fig. (3). T.S of *Channa punctatus* control skin tissue exhibiting the presence of lipids (Sudan black B).

Fig. (4). T.S of *Channa punctatus* infected skin tissue displaying the lipids and phospholipids (Sudan black B).

Ninhydrin/Schiff

The presence of protein-bound amino groups was confirmed by Ninhydrin/Schiff, which yielded a moderately positive response in the control, whereas the EUS-infected skin tissue exhibited a highly positive reaction (Plate **VI**, Figs. **1** to **4**).

PLATE – VI

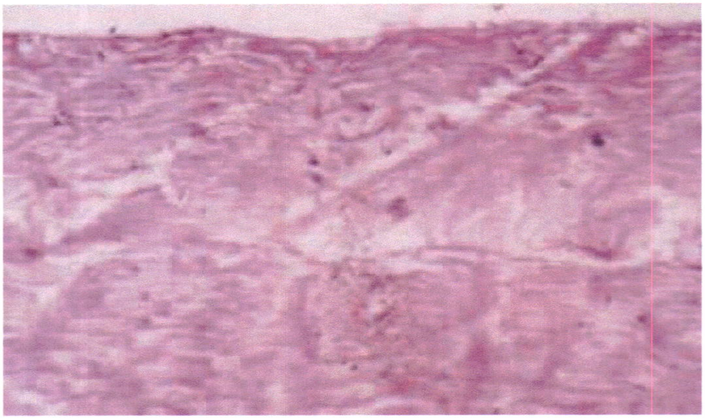

Fig. (1). T.S of *Channa striatus* control skin (Ninhydrin/Schiff).

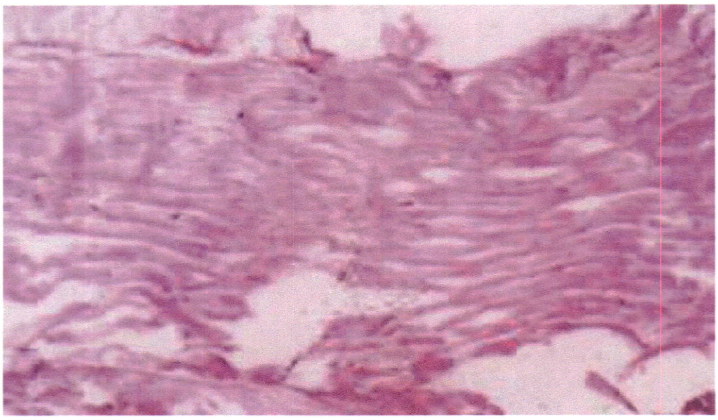

Fig. (2). T.S of *Channa striatus* infected skin (Ninhydrin/Schiff).

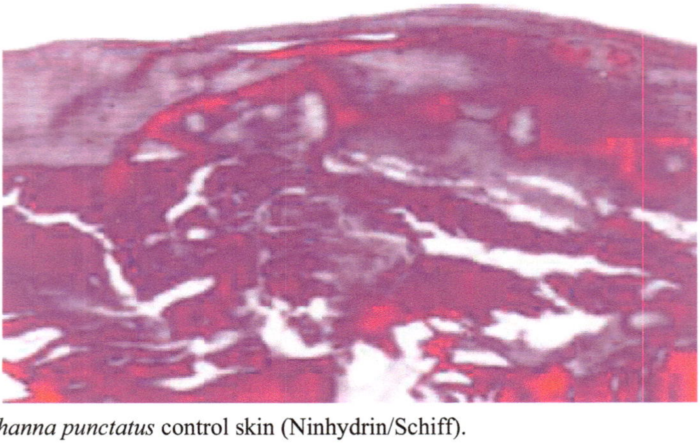

Fig. (3). T.S of *Channa punctatus* control skin (Ninhydrin/Schiff).

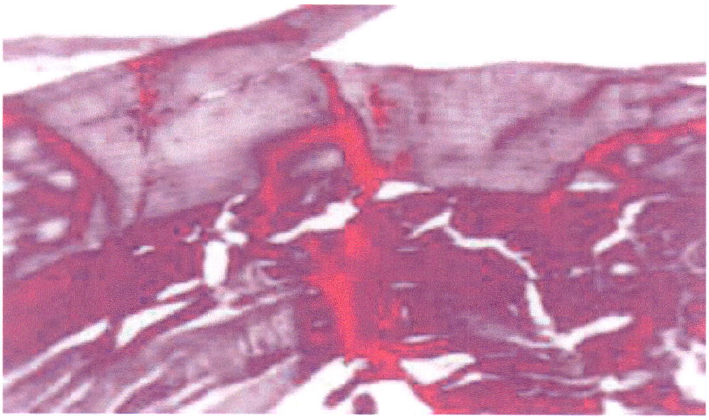

Fig. (4). T.S of *Channa punctatus* infected skin (Ninhydrin/Schiff).

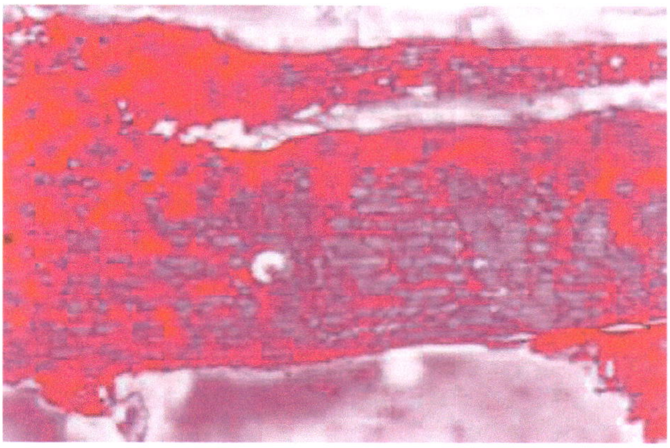

Fig. (5). T.S of *Channa striatus* control skin (Millon's reaction).

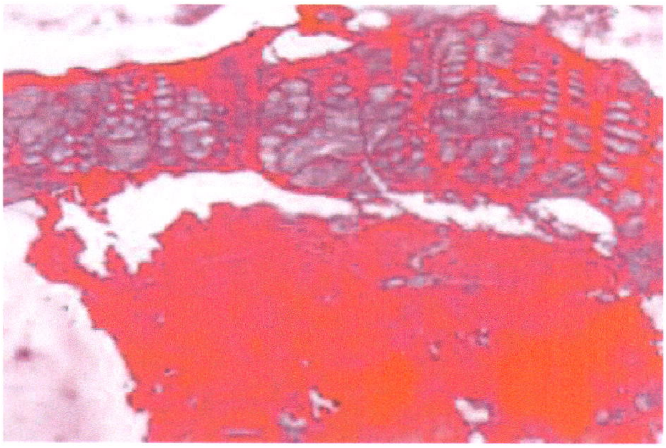

Fig. (6). T.S of *Channa striatus* infected skin (Millon's reaction).

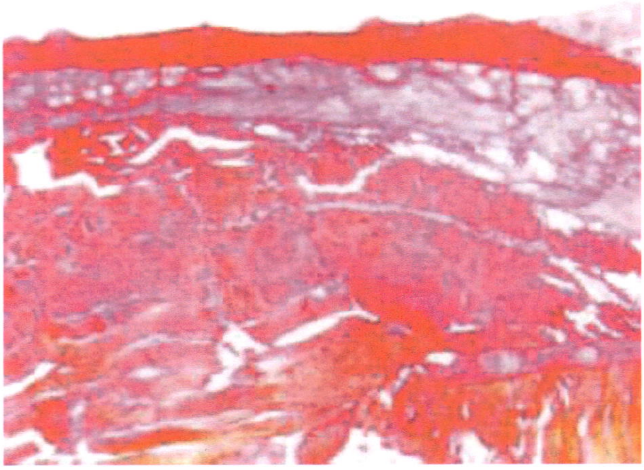

Fig. (7). T.S of *Channa punctatus* control skin (Millon's reaction).

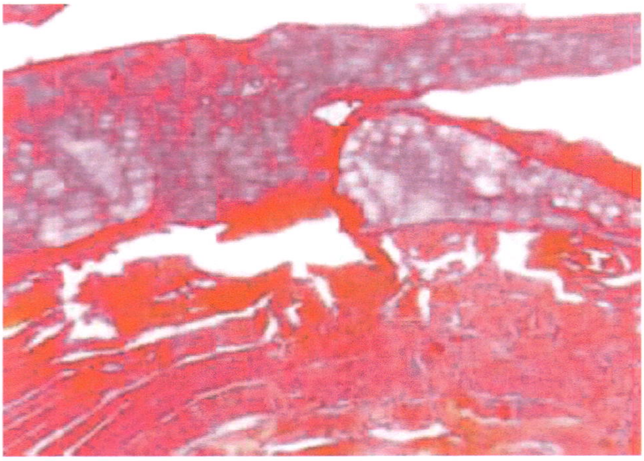

Fig. (8). T.S of *Channa punctatus* infected skin (Millon's reaction).

Millon's Response

Amino acid tyrosine had a weakly positive response in the control skin tissue; however, a significantly positive response was observed in EUS-infected skin using Millon's reaction.

Ferric Ferricyanide (FFC)

The control skin tissue exhibited moderate levels of sulfhydryls and disulfides. In contrast, the EUS, infected skin tissue exhibited a favorable reaction to ferric ferricyanide (Plate **VII**, Figs. **1** to **8**).

PLATE – VII

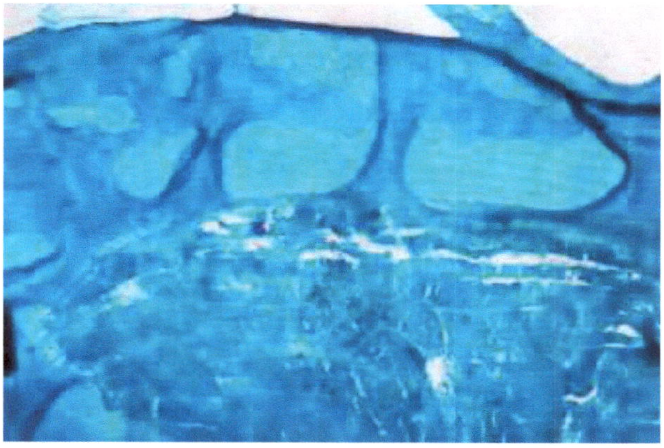

Fig. (1). T.S of *Channa striatus* control skin (FFC).

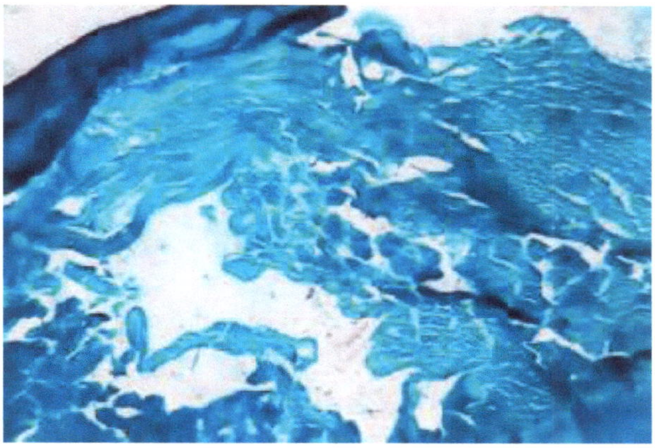

Fig. (2). T.S of *Channa striatus* infected skin (FFC).

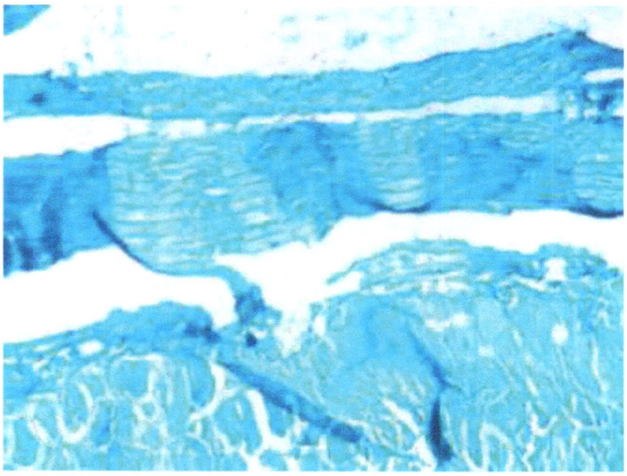

Fig. (3). T.S of *Channa punctatus* control skin (FFC).

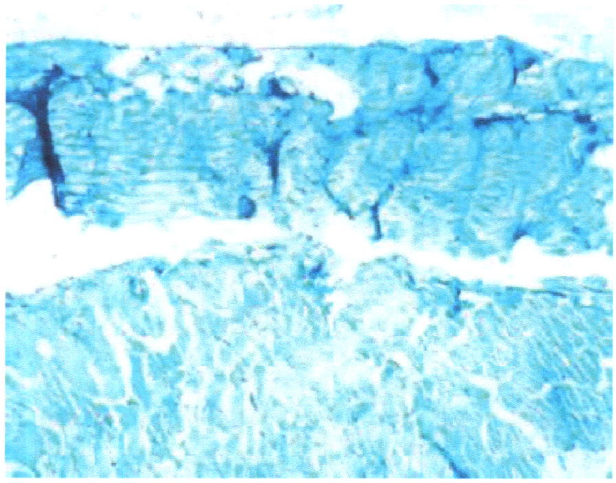

Fig. (4). T.S of *Channa punctatus* infected skin (FFC).

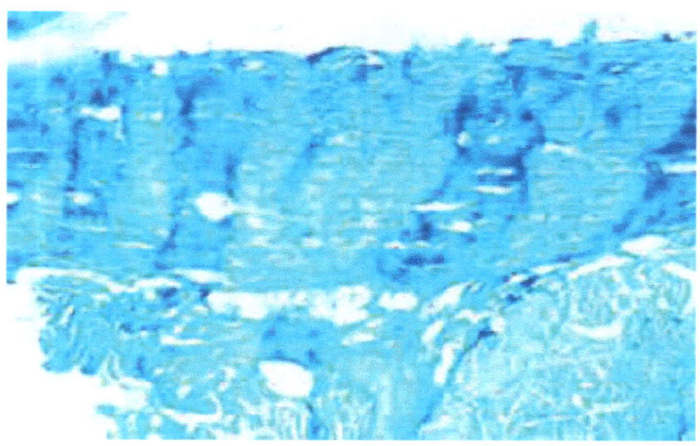

Fig. (5). T.S of *Channa striatus* control skin ($KMnO_4$/Alician Blue).

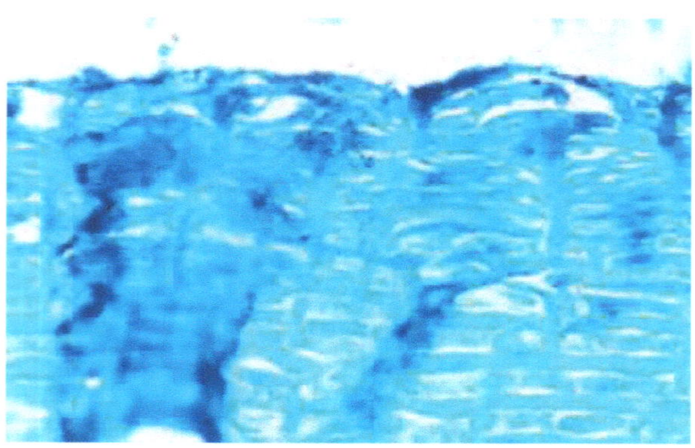

Fig. (6). T.S of *Channa striatus* infected skin ($KMnO_4$/Alician Blue).

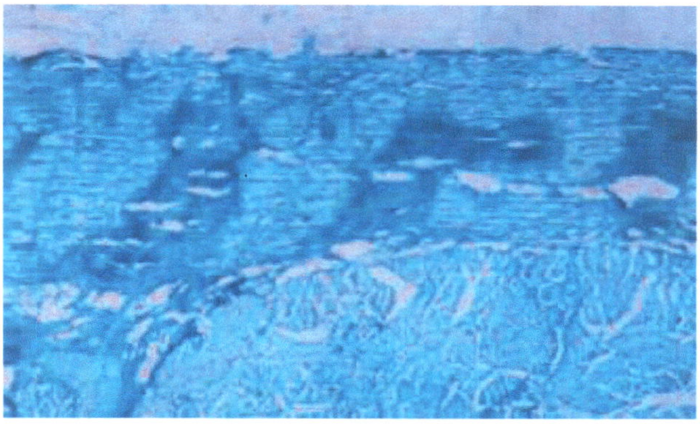

Fig. (7). T.S of *Channa punctatus* control skin ($KMnO_4$/Alician Blue).

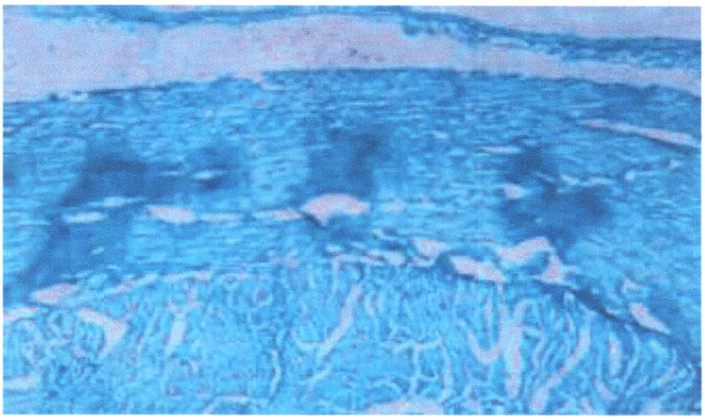

Fig. (8). T.S of *Channa punctatus* infected skin ($KMnO_4$/Alician Blue).

$KMnO_4$/AB

The control skin tissue has a modest positive reaction for cystine with $KMnO_4$/Alcian blue. In the EUS, diseased skin tissue exhibits a positive reaction to $KMnO_4$/Alcian blue.

Copper Phthalocyanine (Cupc)

The skin serves as the primary barrier against the environment, facilitating proper internal physiological functions. The mucous membrane of the skin contains substantial quantities of proteins and phospholipids. The histochemical assays performed on the skin of *C. Striatus* and *C. Punctatus* demonstrated the existence of substantial quantities of phospholipids that exhibited high positivity in control skin tissue when stained with copper phthalocyanine. In the EUS-infected skin tissue, there is a moderate positive reaction to copper phthalocyanine. Consequently, the presence of numerous protein types is observed in the skin mucous cells, reflecting the dynamic activity of proteins (Plate **VIII**, Figs. **1** to **4**).

PLATE – VIII

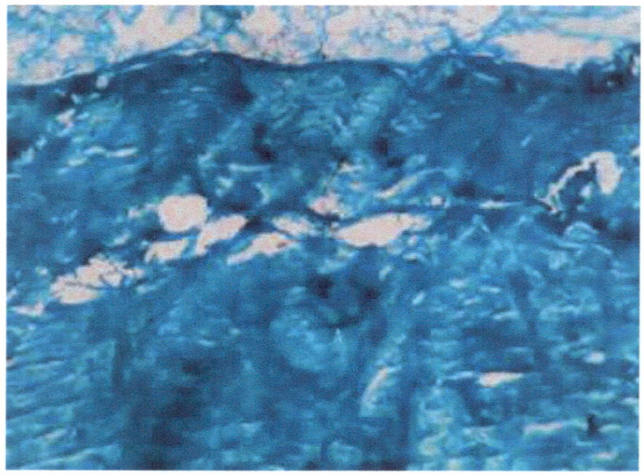

Fig. (1). T.S of *Channa striatus* control skin (CPP).

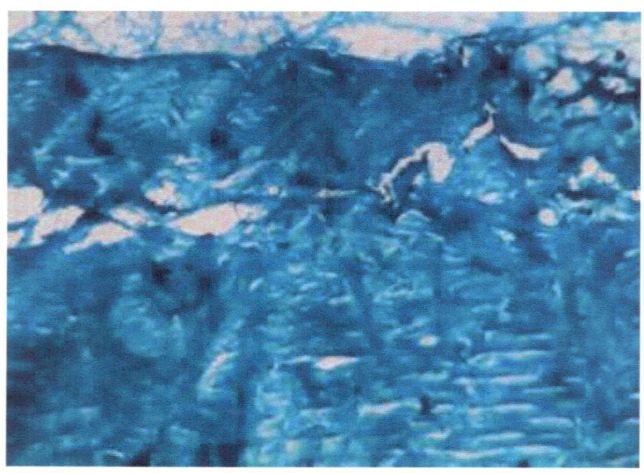

Fig. (2). T.S of *Channa striatus* infected skin (CPP).

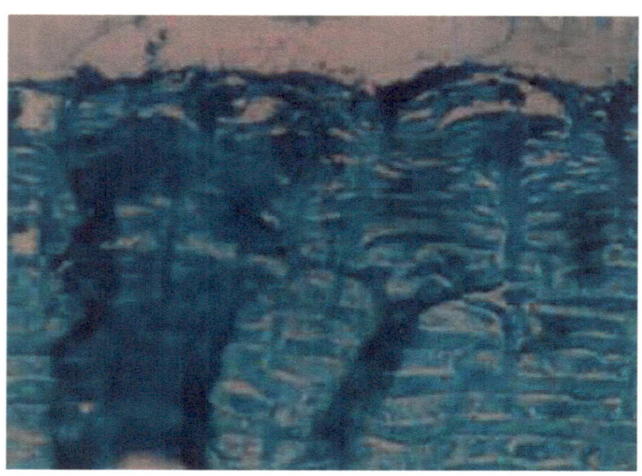

Fig. (3). T.S of *Channa punctatus* control skin (CPP).

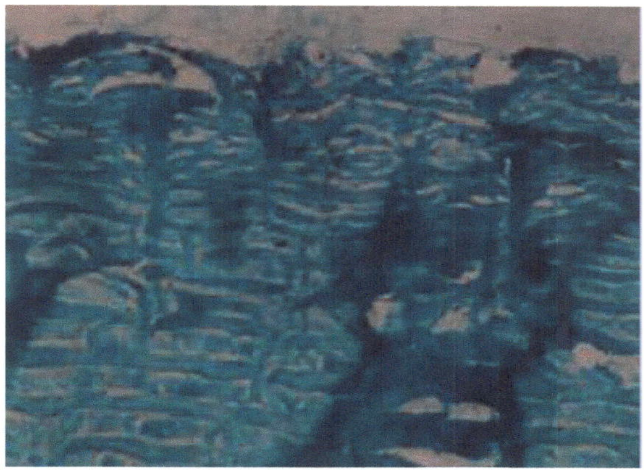

Fig. (4). T.S of *Channa punctatus* infected skin (CPP).

Nucleic Acids

Pyronin y

The skin tissue exhibits robust reactions to the Pyronin Y stain, indicating the presence of extracellular RNA in the control tissue sections.

When the epidermis infected by bacteria and fungi was treated with Pyronin Y, the tissue exhibited a mild reactivity, indicating a depletion of RNA nucleic acids. Flat cells and goblet cells displayed a negative reaction, but bundles of coarse collagenous fibers reacted to Pyronin Y (Plate **IX**, Figs. **1** to **4**).

PLATE – IX

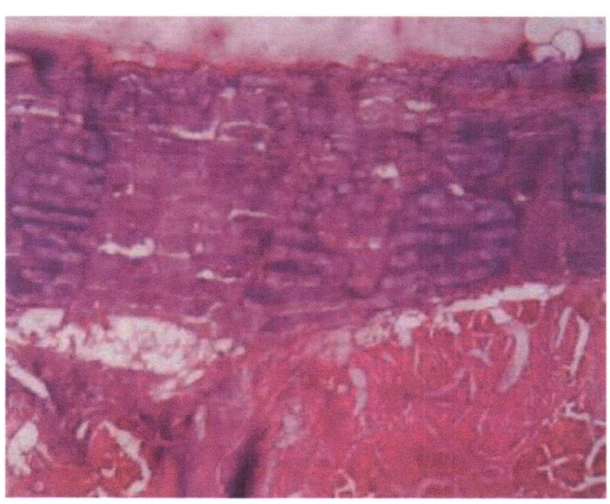

Fig. (1). T.S of *Channa striatus* control skin (Pyronin Y).

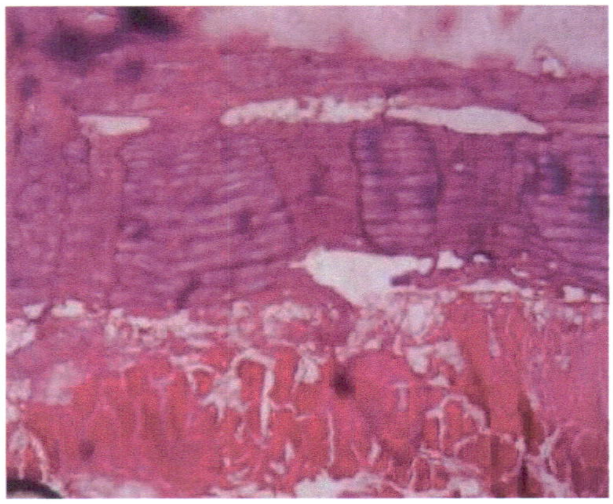

Fig. (2). T.S of *Channa striatus* infected skin (Pyronin Y).

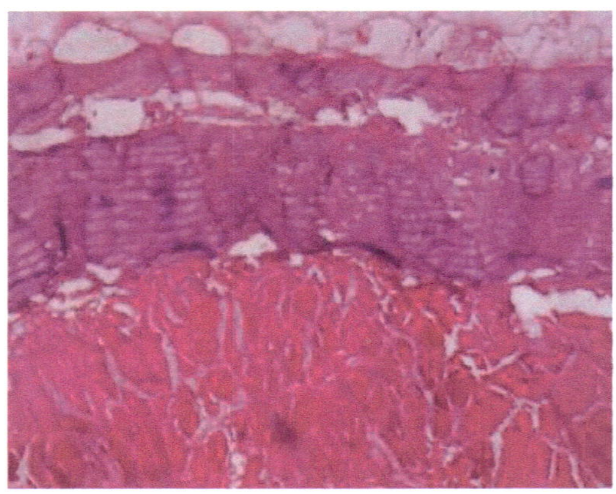

Fig. (3). T.S of *Channa punctatus* control skin (Pyronin Y).

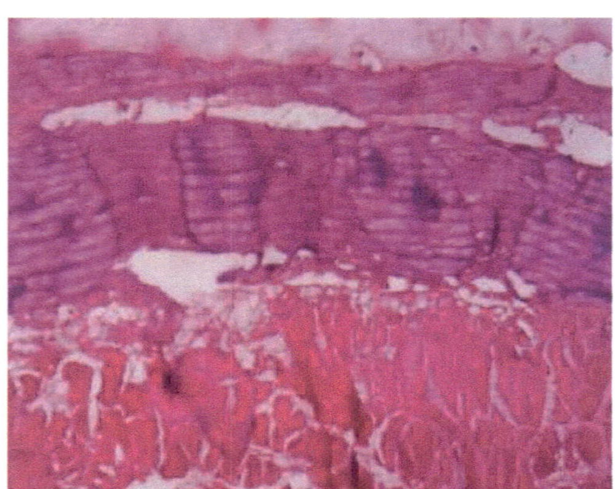

Fig. (4). T.S of *Channa punctatus* infected skin (Pyronin Y).

Feulgen Reaction

The nuclei of epithelial cells are typically centrally located, elongated or flattened, and possess unique chromatin and nucleoli. Flat cells and goblet cells exhibited a positive reaction, but the bundles of coarse collagenous fibers reacted strongly to the Feulgen reaction, indicating the presence of extracellular DNA in the control tissue sections.

When the epidermis was infected by bacteria and fungi, which underwent the Feulgen reaction, it exhibited moderate positive results, indicating a depletion of DNA nucleic acids. The tissue was entirely devoid of proteins, with remains primarily contained within the collagenous fibers (Plate **X**, Figs. **1** to **4**).

PLATE – X

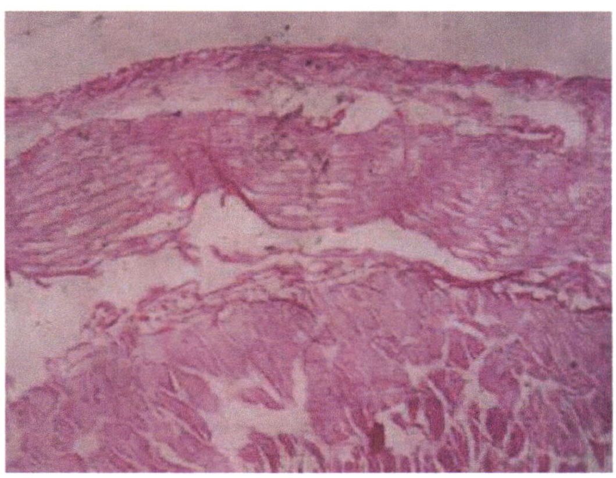

Fig. (1). T.S of control skin (Feulgen reaction).

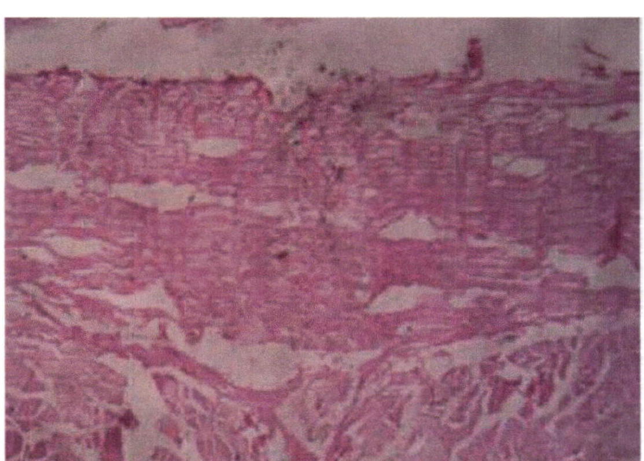

Fig. (2). T.S of *Channa striatus* infected skin (Feulgen Reaction).

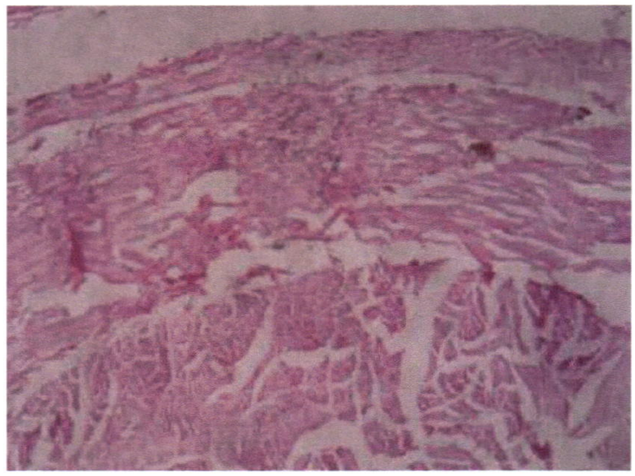

Fig. (3). T.S of *Channa punctatus* control skin (Feulgen Reaction).

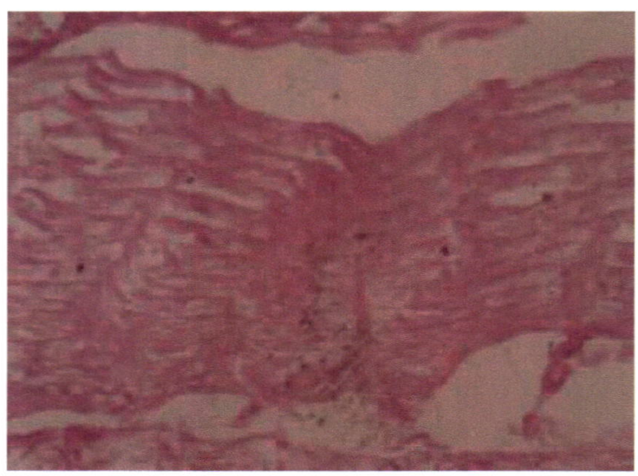

Fig. (4). T.S of *Channa punctatus* infected skin (Feulgen Reaction).

Comparative Histochemical Examination of Control and Diseased Gills

The current investigation of histochemical testing demonstrates that the regulation of freshwater fish of *C. Striatus* and *C. Punctatus* possesses four pairs of gills. Each gill filament or primary gill lamella possesses a series of alternately organized respiratory (secondary) lamellae on either side. Secondary lamellae consist of alternately arranged blood channels and supporting pillar cells, resulting in a ladder-like arrangement. A slender layer of respiratory epithelium envelops the pillar cells, while blood channels constitute the elements of the secondary lamellae. Typically, one or two red blood cells can traverse each blood channel. Mucous cells are predominantly located in the primary lamellae (Singh *et al.*, 2014). The periphery of these mucous cells exhibited moderate staining with PAS and moderate to strong staining with AB 2.5 and AB 1.0, resulting in bluish-violet coloration with AB 2.5/PAS. The thin mucous layer, when present on the primary and secondary lamellae, stains weakly to moderately with AB 2.5 and AB 2.5/PAS, while exhibiting negative staining with AB 1.0 and PAS.

The infected gill blood cells of the secondary lamellae exhibited significant congestion. The augmented weight of these RBCs induced stretching of the respiratory epithelium, resulting in damage that frequently culminated in significant hemorrhage. The pillar cells were vertically compressed due to blood cell congestion. Hemorrhage from the blood cells of secondary lamellae terminated due to bacterial and fungal infection. The gill filaments became densely produced due to significant hyperplasia of the epithelial cells in the main and secondary lamellae when the individual structure of the secondary lamellae was compromised in specific areas. The ladder-like configuration of the blood cells and pillar cells began to distort as the space occupied by the blood cells diminished. This is succeeded by a partial restoration of the ladder-like configuration of the secondary lamellae, despite the blood cell volume remaining markedly reduced and the pillar cells approaching one another significantly.

The compromised secondary lamellae exhibited detachment from the vascular structures, resulting in hemorrhages. Subsequent hyperplasia resulted in the complete embedding of the secondary lamellae within the major lamellae, rendering them solid. Subsequent to the seasonal infectious circumstances, the mucous cells exhibited hyperplasia, succeeded by hypertrophy, resulting in a coating of mucus covering the respiratory surface. The ensuing hyperplasia of epithelial cells resulted in the union of adjacent primary lamellae in the later stages. Despite the ongoing recurrent proliferation of the chloride cells, they often underwent degeneration. The density, dimensions, secretory activity, and staining characteristics of the mucous cells varied independently during various stages of infection. The amalgamation of neighboring main lamellae persisted despite seasonal recuperation. The aberrant histo-morphology of the primary lamellae, coupled with destroyed blood cells and components of the pillar cells in the secondary lamellae, persisted in multiple locations. Red blood cells were consistently found among the dispersed blood cells during infection, accompanied by significant healing of the gill filaments. The ladder-like vascular structure of the secondary lamellae was reformed with significantly reduced thickness of the epithelial lining. The gill regeneration persisted through the next stages. Nevertheless, in the inner layers, these cells exhibit a loosely woven structure. At specific later stages, the small mucous cells predominantly located in the outer epithelial layer have a greater affinity for Alcian Blue (AB) staining. Unidentified cell masses or degenerated mucous cells exhibiting significant staining with Aldehyde fuchsine (AF) were also observed in the deeper layer of the main lamellae.

Glycogen Concentration

Periodic Acid-Schiff (PAS) Reaction

This stain exhibited moderate to strong PAS positivity in control gills within mucous cells, observed in primary or secondary lamellae. The blood cells in the secondary lamellae became significantly engorged with red blood cells that stained positively using the PAS technique.

In the EUS-infected gills, substantial intercellular vacuolization and widespread hyperplasia of the epithelial cells in the primary and secondary lamellae led to increased thickness due to the infection. The detachment of the epithelial lining from both the primary and secondary lamellae was frequently observed during infection. The density of mucous cells in both the primary and secondary lamellae is significantly elevated. These cells were stained negatively with Periodic Acid-Schiff (PAS).

Alcian Blue (Ab) pH 2.5

The AB 2.5 stain exhibited moderate to high affinity with control gills at this stage; a considerable number of rounds and big vacuoles of uniform size commenced forming in both the main and secondary lamellae. A minimal quantity of basophilic, viscous material had a favorable reaction with AB 2.5.

The gills infected with bacteria and fungi exhibited a modest reduction in mucous cell density, and a basophilic, fuzzy substance was consistently shed from the mucus-coated surfaces of the primary and secondary lamellae, which tested negative with AB 2.5.

Alcian Blue (Ab) pH 1.0

In the control gills, vacuolization intensified, with numerous big vacuoles of uniform size emerging in both the primary and secondary lamellae. A little quantity of basophilic, viscous material stained positively with AB 1.0 was commonly noted, particularly in the inner lining of these vacuoles. The dimensions and density of vacuoles in the hyperplastic primary lamellae, particularly in the secondary lamellae, escalated in the later stages. Although minor degradation of the epithelial linings of both the primary and secondary lamellae was observed at this stage, no hemorrhage or rupture of blood cells was identified.

The gills infected by bacteria and fungi exhibit significantly dilated and engorged blood cells in the principal lamellae post-infection. Additionally, the density of mucous cells is diminished. A mildly basophilic, fuzzy material is consistently shed from the surface of the secondary lamellae. A thin layer of mucus covered the surface of the secondary lamellae, resulting in the collapse of multiple spots where the typical ladder-like arrangement of pillar cells and blood cells is observed. Nonetheless, the gills partially regenerated and restored some of their diminished staining characteristics at several other locations. The blood capillaries within the gill filament exhibited significant congestion and engorgement with many red blood cells (Plate **XI**; Figs. **1** to **12**).

PLATE – XI

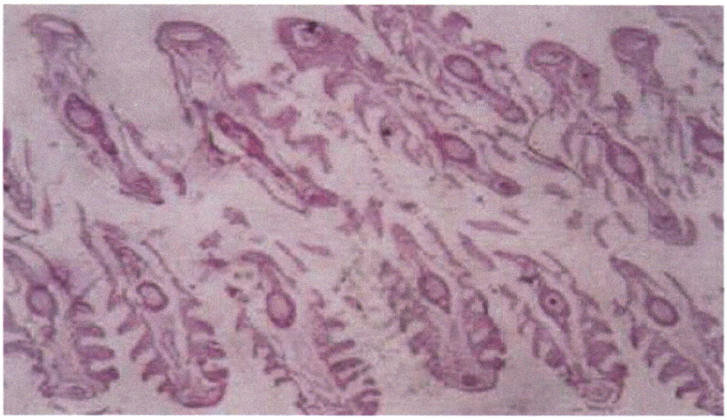

Fig. (1). T.S of *Channa striatus* control gill shows moderate presence of Glycogen (PAS).

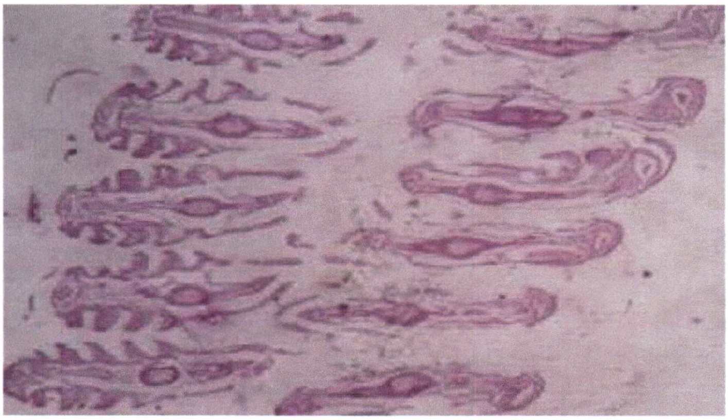

Fig. (2). T.S of *Channa striatus* infected gill shows a strong presence of glycogen substances (PAS).

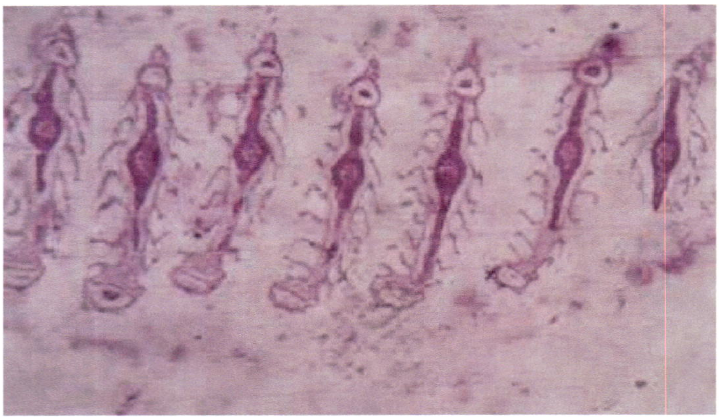

Fig. (3). T. S of *Channa striatus* control gill showing pale blue presence of acid mucin (AB2.5pH).

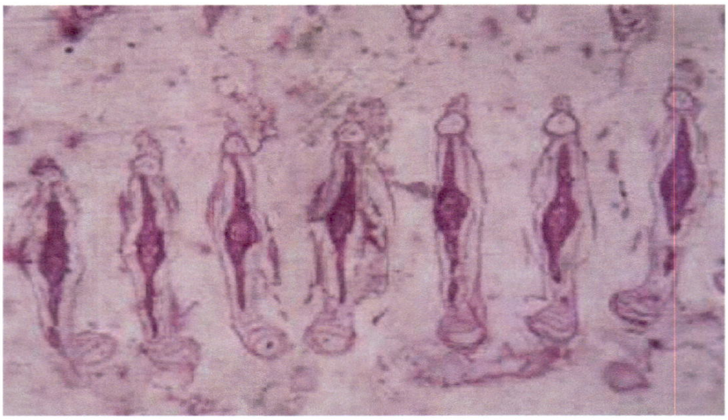

Fig. (4). T.S of *Channa striatus* infected gill shows light and dark blue presence of sulphated mucins (AB 2.5 pH).

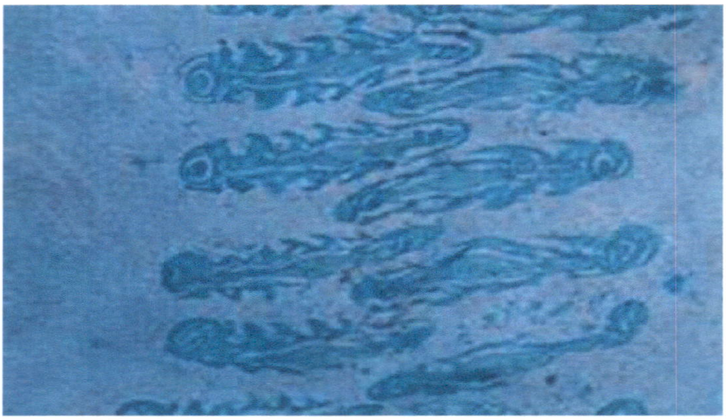

Fig. (5). T.S of *Channa striatus* control gill shows the presence of mucin (AB 1.0 pH).

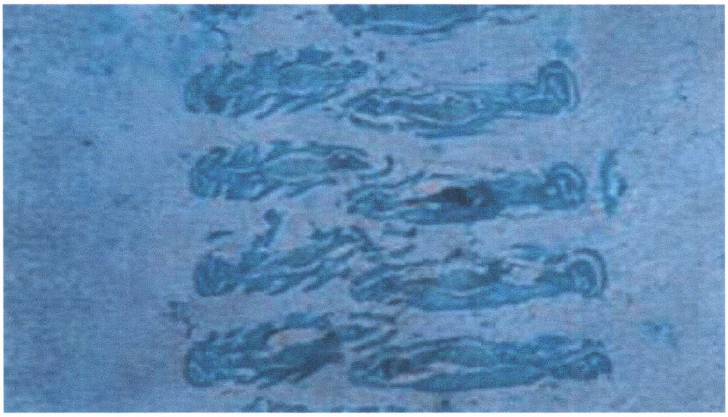

Fig. (6). T.S of *Channa striatus* infected gill showing presence of sulfated mucosubstances (AB1.0 pH).

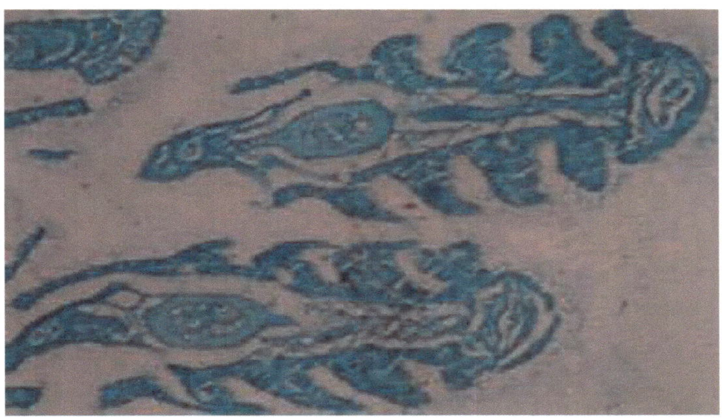

Fig. (7). T.S of *Channa punctatus* control gill showing moderate presence of glycogen (PAS).

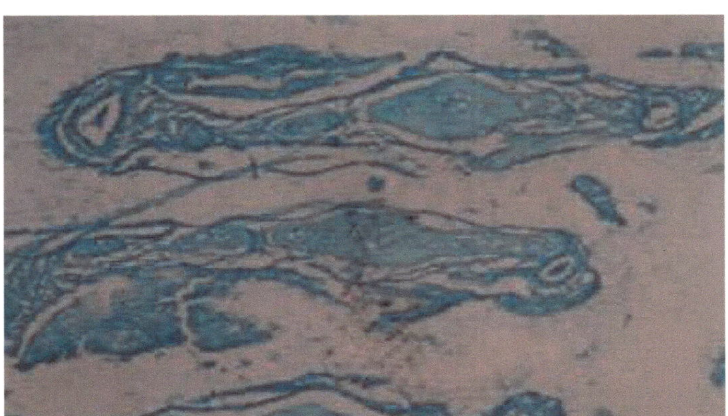

Fig. (8). T.S of *Channa punctatus* infected gill showing the strong presence of glycogen substances (PAS).

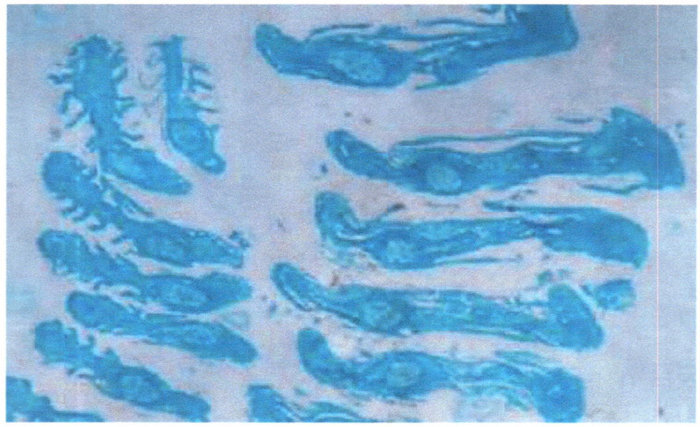

Fig. (9). T.S of *Channa punctatus* control gill showing pale blue presence of acid mucin (AB2.5pH).

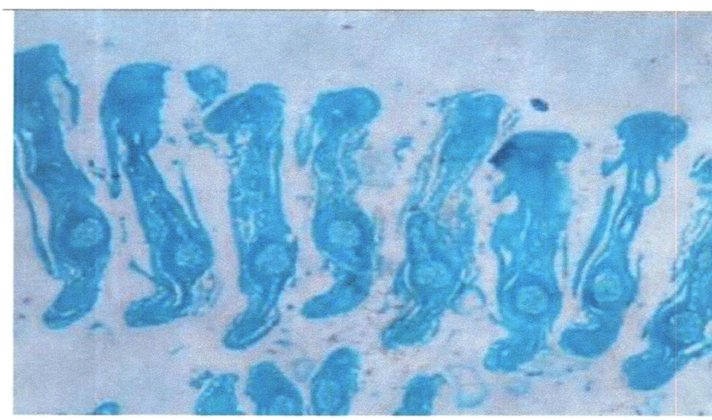

Fig. (10). T.S of *Channa punctatus* infected gill showing light and dark blue presence of sulphated mucins (AB 2.5 pH).

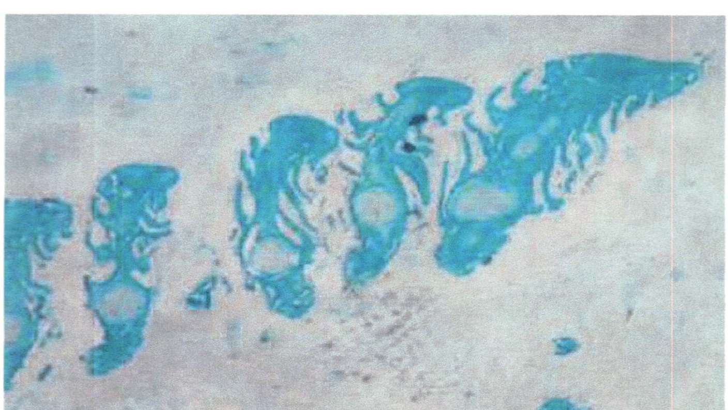

Fig. (11). T.S of *Channa punctatus* control gill showing presence of mucin (AB 1.0 pH).

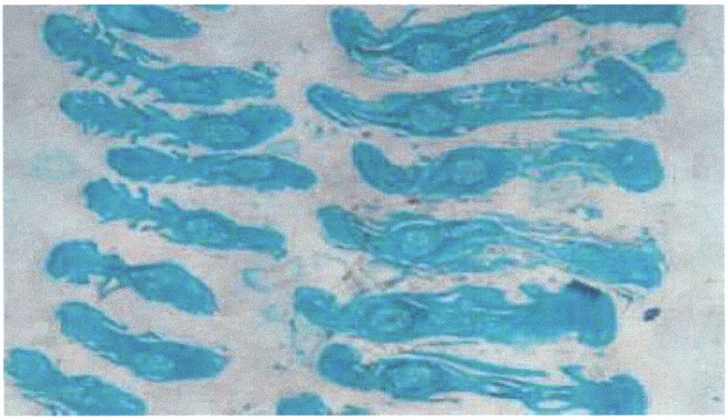

Fig. (12). T.S of *Channa punctatus* infected gill showing presence of sulfated mucosubstances (ABl.0 pH).

Alcian Blue (Ab) pH 2.5/PAS

The control gill demonstrates that the vascular elements of the secondary lamellae consist of alternately organized pillar cells and blood channels, which are enveloped by a thin respiratory epithelium. Mucous cells are predominantly found in the epithelium of the gill filaments or main lamellae. A few saucer-shaped mucous cells are also located in the secondary lamellae. The mucous cells exhibit a dark greenish-black hue with AB 2.5/PAS. The mucous cells on the major lamellae stain pale greenish-blue, while their periphery displays a dark blackish-green color with AB 2.5/PAS.

EUS-infected *C. Striatus* and *C. Punctatus* gills exhibited extensive intercellular vacuolization accompanied by widespread hyperplasia of the epithelial cells in both primary and secondary lamellae, leading to increased thickness. The density of mucous cells in both lamellae significantly increased, displaying a greenish-blue hue when analyzed using AB 2.5/PAS techniques.

Aldehyde Fuchsine

The aldehyde fuchsine staining of normal gills revealed vascular structures within the secondary lamellae in purple, while blood channels were obscured by a thin layer of respiratory epithelium stained pale yellow; pillar cells exhibited a negative reaction. The epithelial cells of the gills exhibit a vivid purple hue. The crimson coloration of the mucous cells indicated that they were likely sulfomucins.

The EUS-infected gill lamellae tissues had a high positive reaction to Aldehyde fuchsine, indicating the presence of sulfated mucosubstances. Conversely, the infected gill tissue indicated a reduction in the epithelium (Plate **XII**; Figs. **1** to **8**).

PLATE – XII

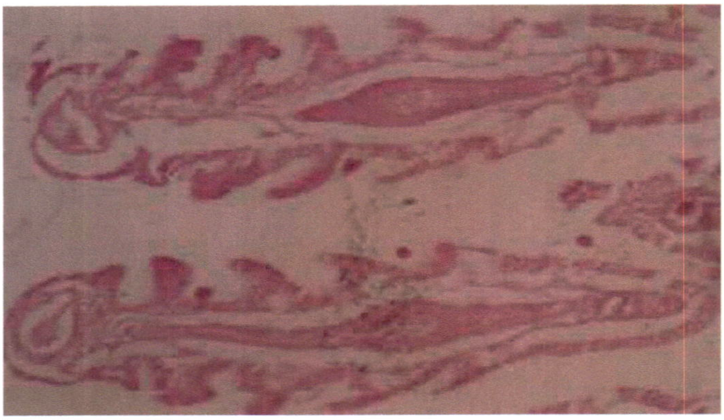

Fig. (1). T.S of *Channa striatus* control gill, displaying the presence of glycogen predominantly pink, a lesser amount of mucin blue (AB25pH/PAS).

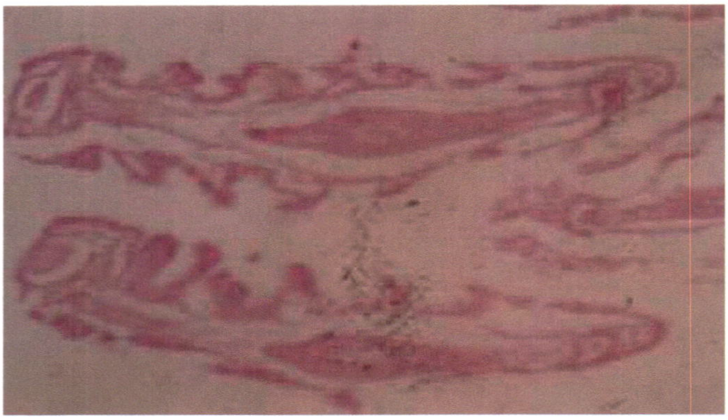

Fig. (2). T.S of *Channa striatus* infected gill displaying the presence of mucin predominantly (AB 2.5 pH /PAS).

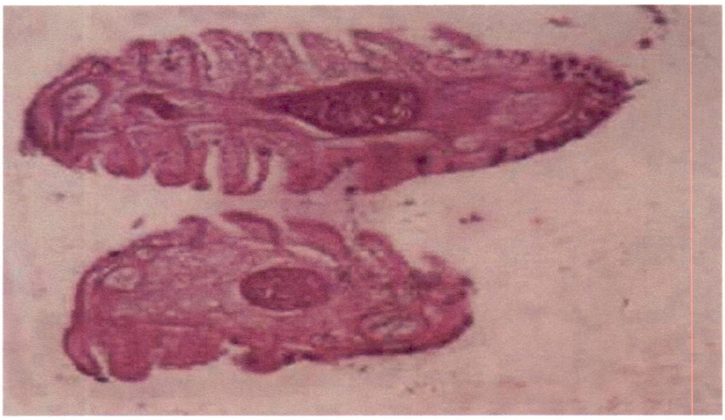

Fig. (3). T.S of *Channa striatus* control gill presented elastic fibers stained purple, and collagen stained with light yellow (Aldehyde fuchsine).

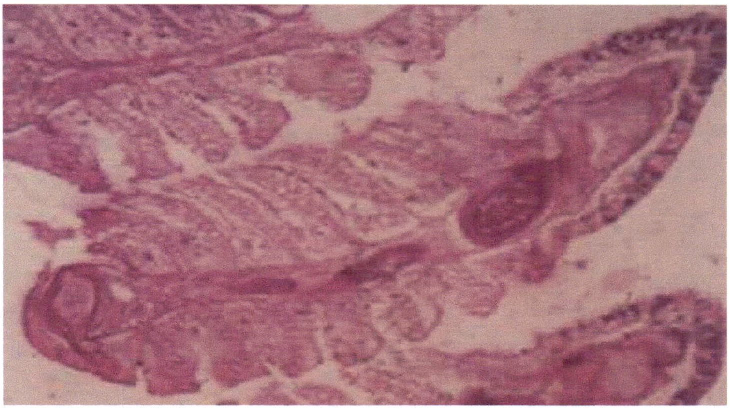

Fig. (4). T.S of *Channa striatus* infected gill showing decreased Aldehyde fuchsine.

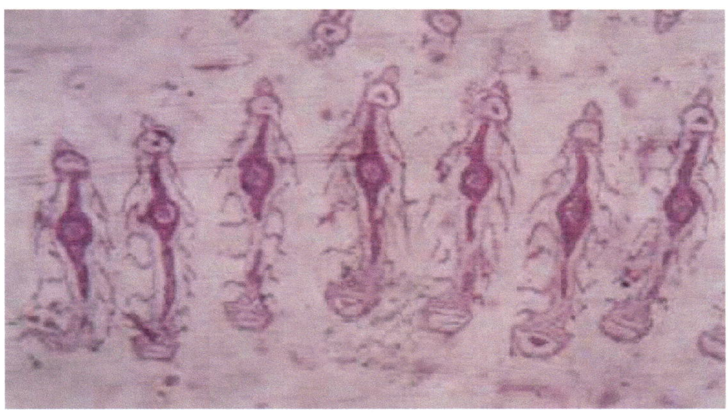

Fig. (5). T.S of *Channa punctatus* control gill, displaying the presence of glycogen predominantly (pink), a lesser amount of mucin (blue) (AB25pH/PAS).

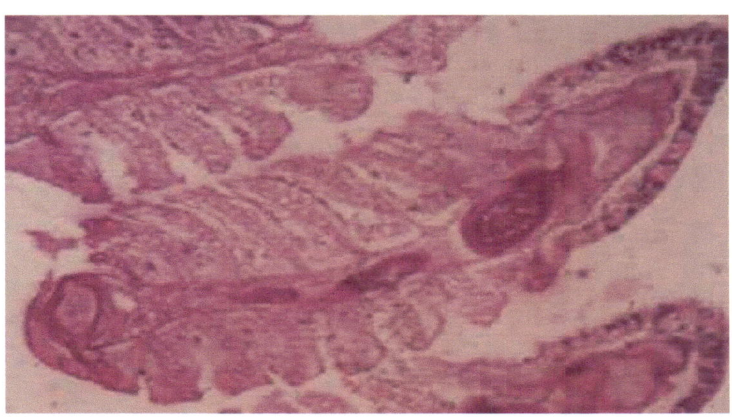

Fig. (6). T.S of *Channa punctatus* infected gill displaying the presence of mucin predominantly (AB 2.5 pH /PAS).

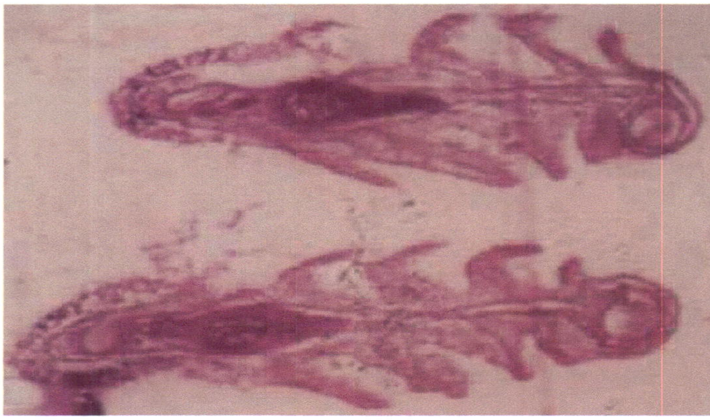

Fig. (7). T.S of *Channa punctatus* control gill presented elastic fibers, purple, and collagen stained with light yellow (Aldehyde fuchsine).

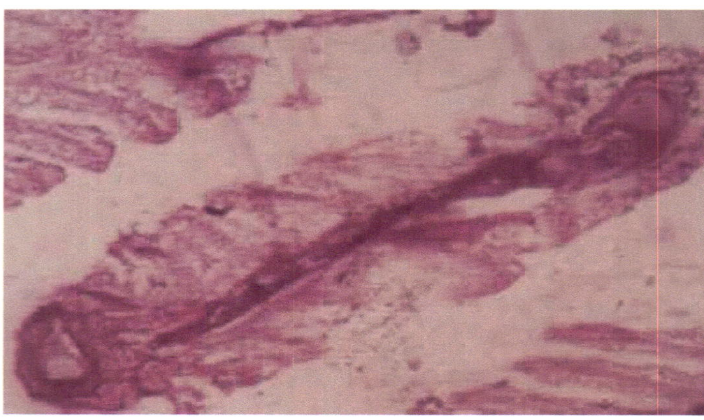

Fig. (8). T.S of *Channa punctatus* infected gill showing decreased (Aldehyde fuchsine).

Proteins

Mercuric Bromophenol Blue

The epithelial cells of control gill lamellae envelop the pillar cell system, collectively constituting the respiratory surface area of the gill lamellae. These are referred to as respiratory epithelial cells. The cytoplasmic contents of these pillar cells exhibit a pronounced bluish staining, while the cytoplasm of blood channels displays a weaker stain. The cell peripheries and nuclei show strong staining, whereas the nuclei and cell peripheries demonstrate a moderate reaction with the Hg-BPB method for general and basic proteins.

The epithelial cells of the gill lamellae, infected by bacteria and fungi, envelop the pillar cells, together constituting the respiratory surface area of the gill lamellae. These are referred to as respiratory epithelial cells. The cytoplasmic contents exhibit a modest reaction, whereas the nuclei and cell peripheries demonstrate a moderate reaction using the Hg-BPB technique for general and basic proteins.

The histochemical methods employed in this work have distinctly demonstrated that in the two fish species. The gills contain a substantial quantity of proteinaceous material. This was confirmed when the gill sections were submitted to specialized stains for the detection of various proteins (Plate **XIII**; Figs. **1** to **4**).

PLATE – XIII

Fig. (1). T.S of *Channa striatus* control gill showing the presence of basic proteins (Bromophenol blue).

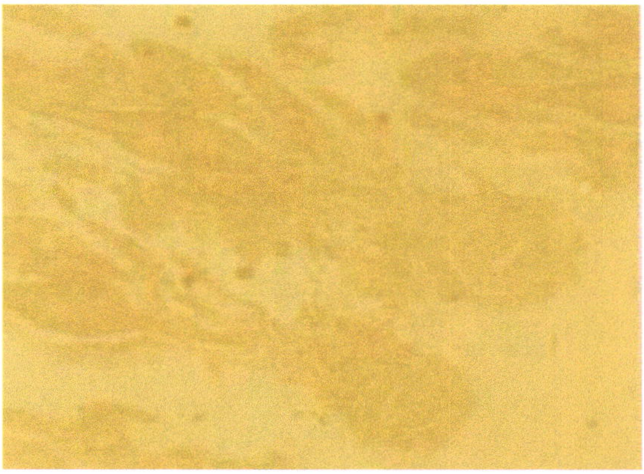

Fig. (2). T.S of *Channa striatus* infected gill showing depletion of basic proteins (Bromophenol blue).

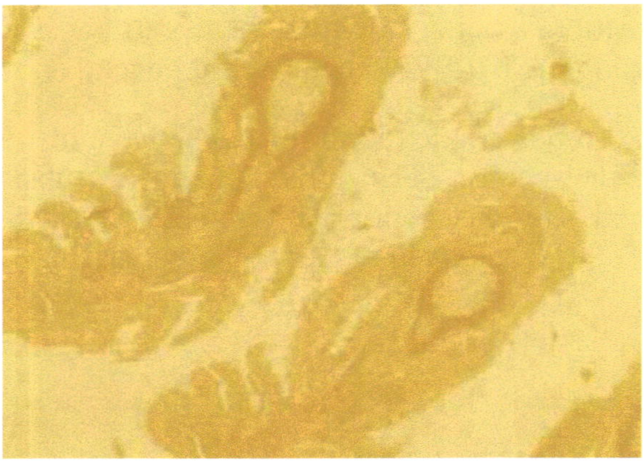

Fig. (3). T.S of *Channa punctatus* control gill showing the presence of basic proteins (Bromophenol blue).

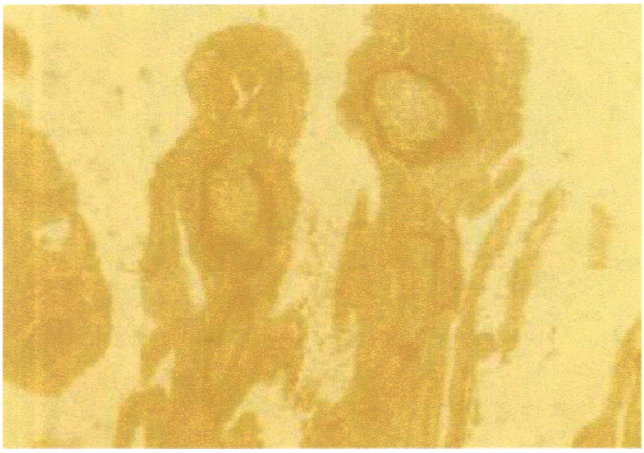

Fig. (4). T.S of *Channa punctatus* infected gill showing depletion of basic proteins (Bromophenol blue).

Congo Red

This approach is employed to illustrate the existence of glycoproteins. A pronounced positive was noted in the control gill tissue. This positive indicated the existence of protein-bound amino groups and glycoproteins within the cells. These results show that glycoproteins are the primary components of epithelial cells. Only the lamellar cells and pillar cells had a pronounced positive effect. This observation, supported by the responses of these cells to other previously utilized stains, indicates the involvement of glycoproteins.

The bacterial and fungal-infected gill tissues, along with the primary lamellae and secondary lamellar epithelial cells, had a somewhat favorable reaction when treated with Congo red. It was entirely devoid of protein, with remains primarily situated in the cytoplasm of blood vessels (Plate **XIV**; Figs. **1** to **4**).

PLATE – XIV

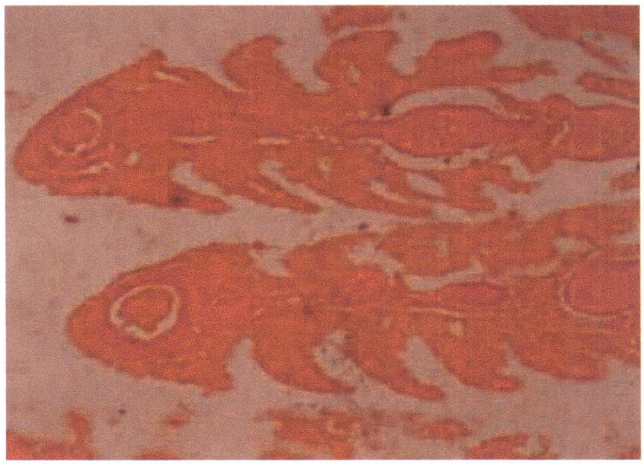

Fig. (1). T.S of *Channa striatus* control gill showing the presence of protein-bound amino groups and glycoproteins in the cells (Congo red).

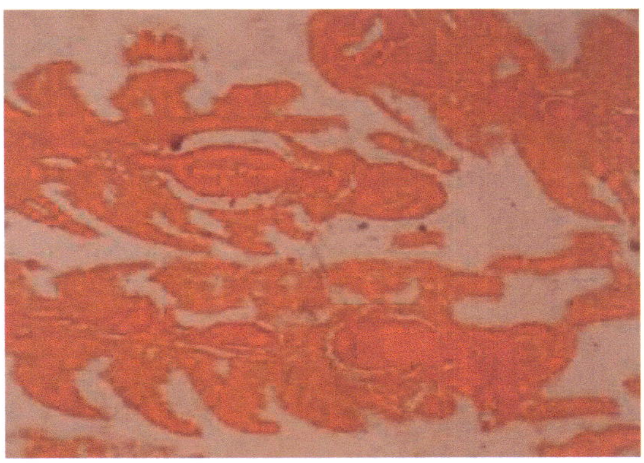

Fig. (2). T.S of *Channa striatus* infected gill showing reduction of glycoproteins (Congo red).

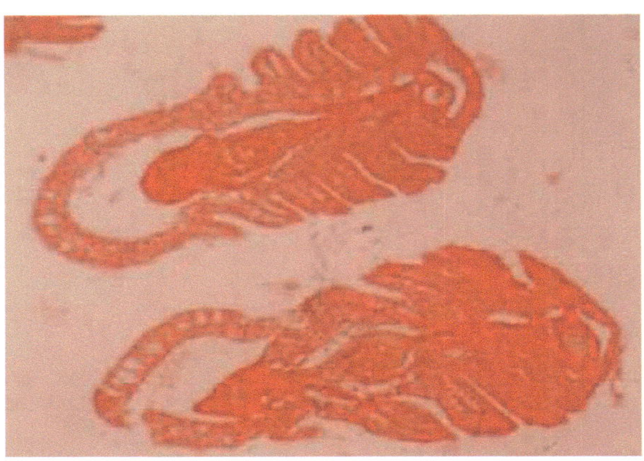

Fig. (3). T.S of *Channa punctatus* control gill showing the presence of protein-bound amino groups and glycoproteins in the cells (Congo red).

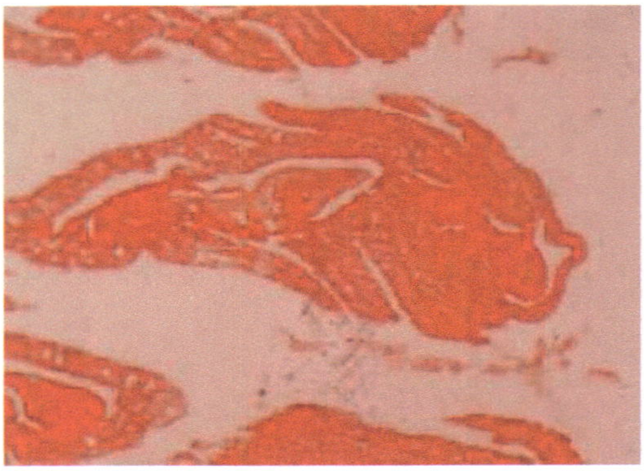

Fig. (4). T.S of *Channa punctatus* infected gill showing reduction of glycoproteins (Congo red).

Fats

Sudan Black B

Calcium-fixed tissue was sectioned and stained with Sudan black B. The control gill tissues of primary and secondary lamellae epithelial cells demonstrated significant positive staining, indicating the presence of lipids. This positive was entirely eliminated when sections underwent pyridine extraction. This additionally corroborated the existence of lipids within the epithelial cells. The pillar cells exhibited positivity to Copperphthalocyanin, indicating the presence of phospholipids.

In the EUS, infected gill tissues were treated with Sudan black B, resulting in a moderate positive reaction in the primary and secondary lamellae epithelium, indicating the presence of lipids and phospholipids (Plate **XV**; Figs. **1** to **4**).

PLATE – XV

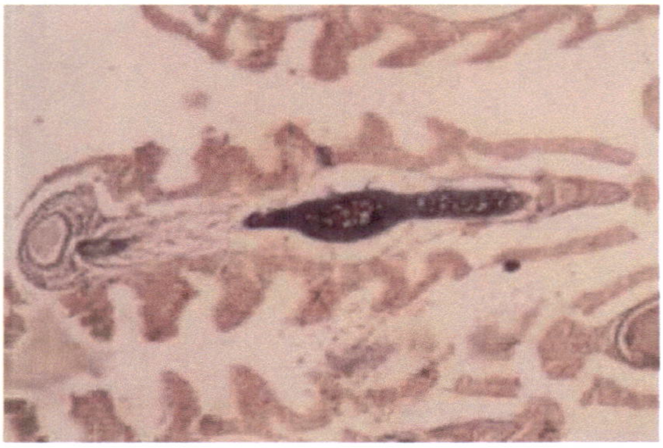

Fig. (1). T.S of *Channa striatus* control gill tissue exhibiting the presence of lipids (Sudan black B).

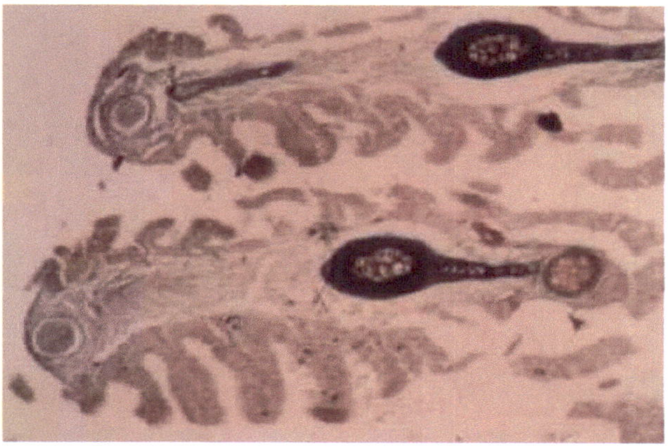

Fig. (2). T.S of *Channa striatus* infected gill tissue displaying the lipids and phospholipids (Sudan black B).

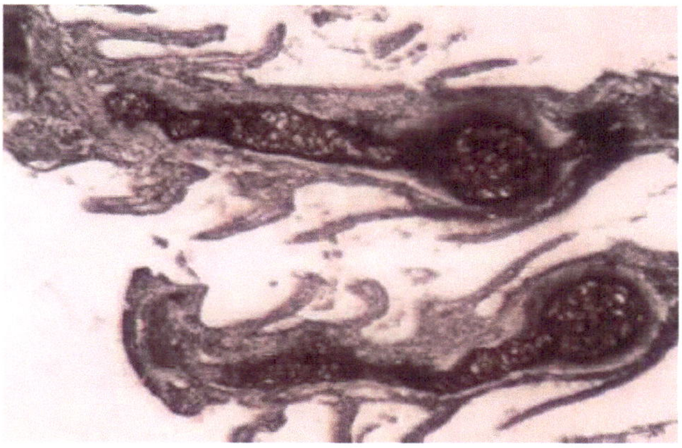

Fig. (3). T.S of *Channa punctatus* control gill tissue exhibiting the presence of lipids (Sudan black B).

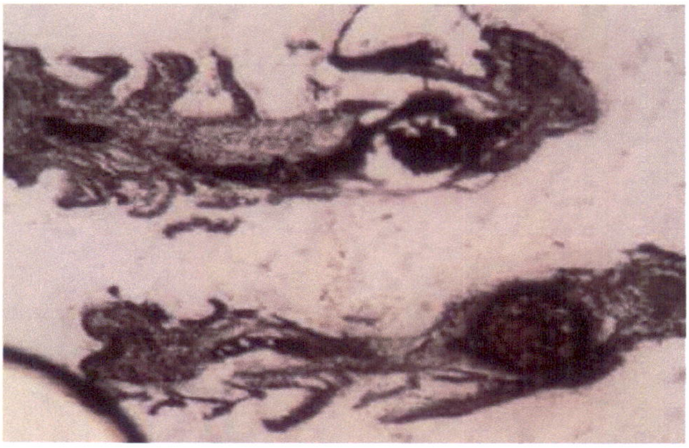

Fig. (4). T.S of *Channa punctatus* infected gill tissue displaying the lipids and phospholipids (Sudan black B).

Ninhydrin/Schiff

The existence of protein-bound amino groups was confirmed using Ninhydrin/Schiff, yielding a moderately positive reaction in the control gill tissue and a substantially positive reaction in the EUS-infected gill tissue (Plate **XVI**, Figs. **1** to **8**).

PLATE – XVI

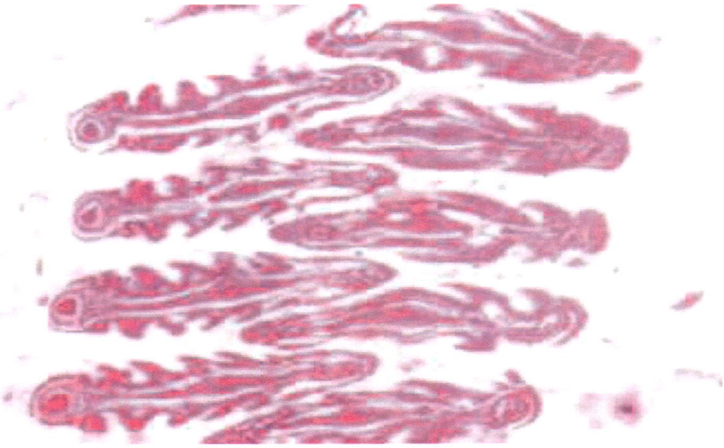

Fig. (1). T.S of *Channa striatus* control gill (Ninhydrin/Schiff).

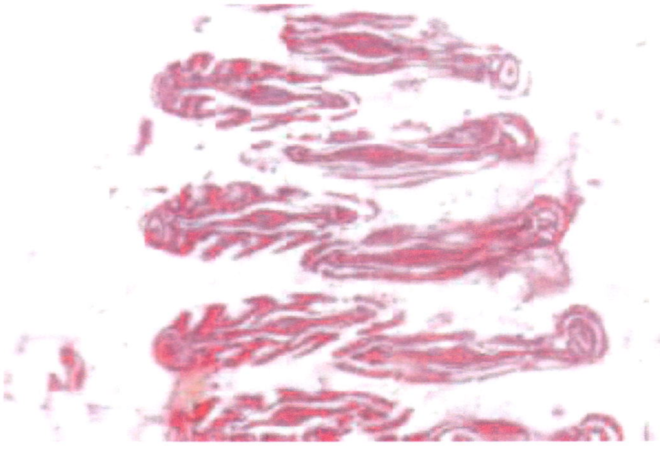

Fig. (2). T.S of *Channa striatus* infected gill (Ninhydrin/Schiff).

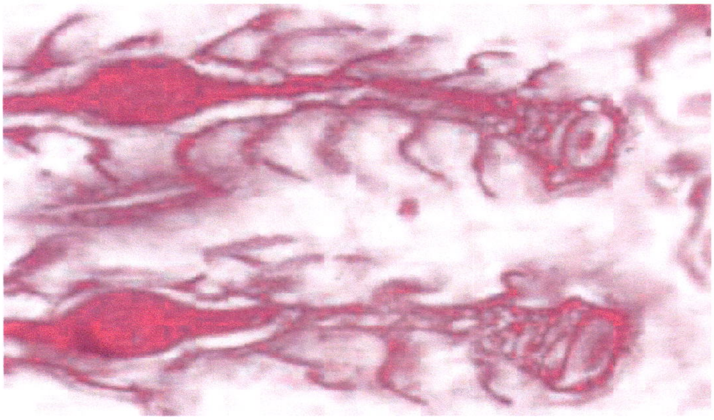

Fig. (3). T.S of *Channa punctatus* control gill (Ninhydrin/Schiff).

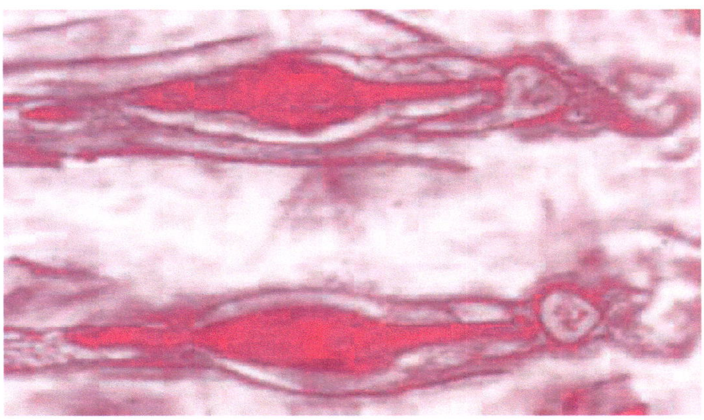

Fig. (4). T.S of *Channa punctatus* infected gill (Ninhydrin/Schiff).

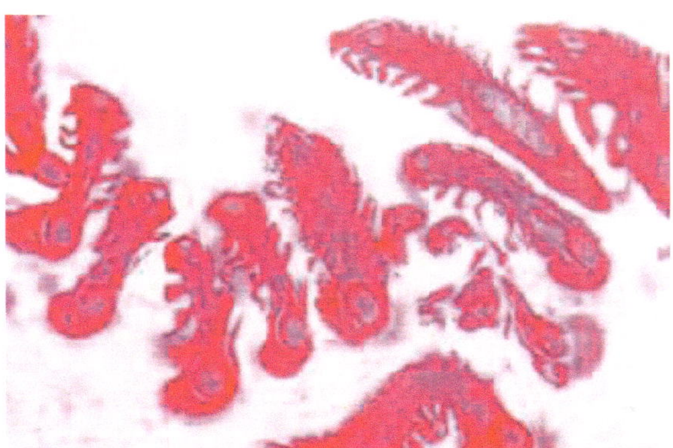

Fig. (5). T.S of *Channa striatus* control gill (Millon's reaction).

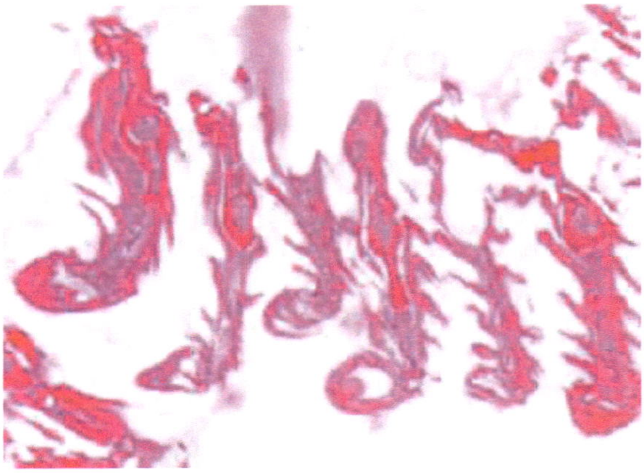

Fig. (6). T.S of *Channa striatus* infected gill (Millon's reaction).

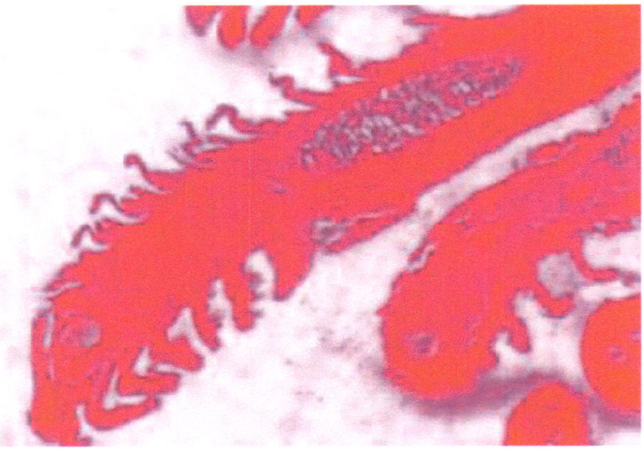

Fig. (7). T.S of *Channa punctatus* control gill (Millon's reaction).

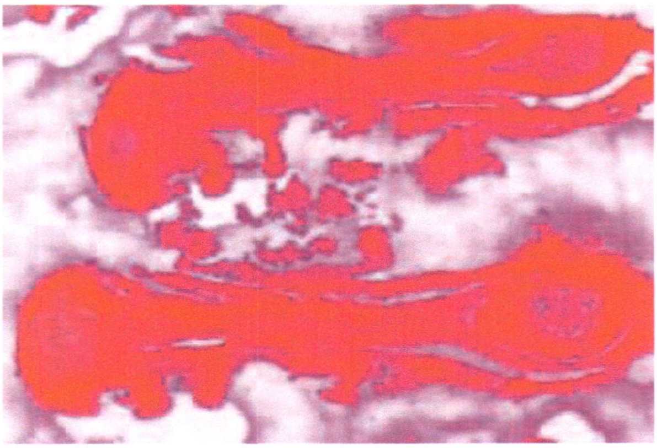

Fig. (8). T.S of *Channa punctatus* infected gill (Millon's reaction).

Millon's Response

Amino acid tyrosine exhibited a somewhat favorable response in the control gill tissue and a positive response in the gill tissue infected with EUS, as indicated by Millon's reaction.

Ferric Ferricyanide (ffc)

The control gill tissue exhibits modest levels of sulfhydryls, and disulfides were also identified. In the EUS-infected gill tissue, a favorable reaction to ferric ferricyanide was observed (Plate **XVII**, Figs. **1** to **4**).

$KMnO_4$/AB

The control gill tissue has a modest positive reaction for cystine with $KmnO_4$/Alcian blue. In the EUS-infected gill tissue, a favorable reaction to $KMnO_4$/Alcian blue was seen (Plate **XVII**, Figs. **5** to **8**.

PLATE - XVII

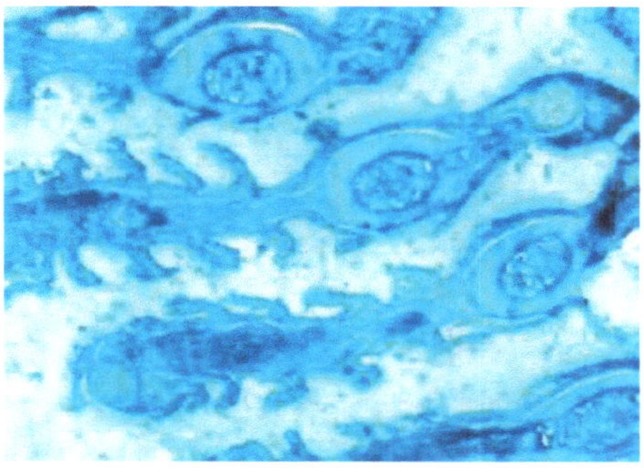

Fig. (1). T.S of *Channa striatus* control gill (FFC).

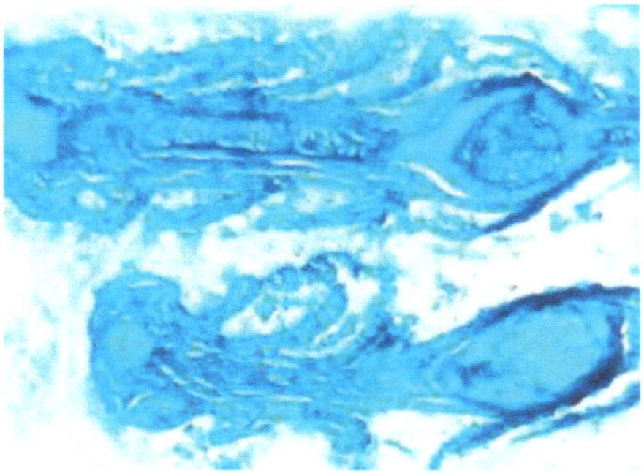

Fig. (2). T.S of *Channa striatus* infected gill (FFC).

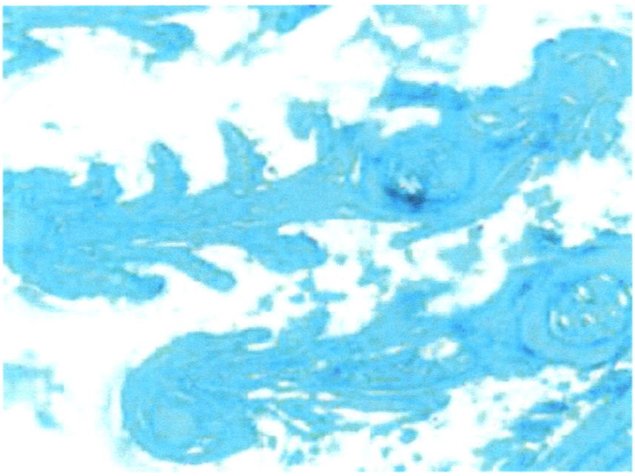

Fig. (3). T.S of *Channa punctatus* control gill (FFC).

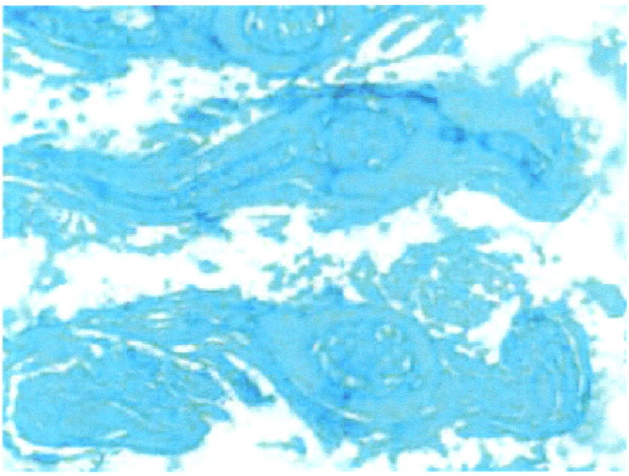

Fig. (4). T.S of *Channa punctatus* infected gill (FFC).

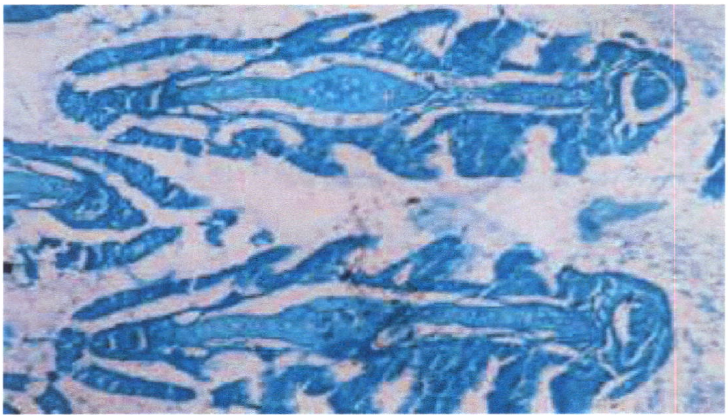

Fig. (5). T.S of *Channa striatus* control gill ($KMnO_4$/Alician Blue).

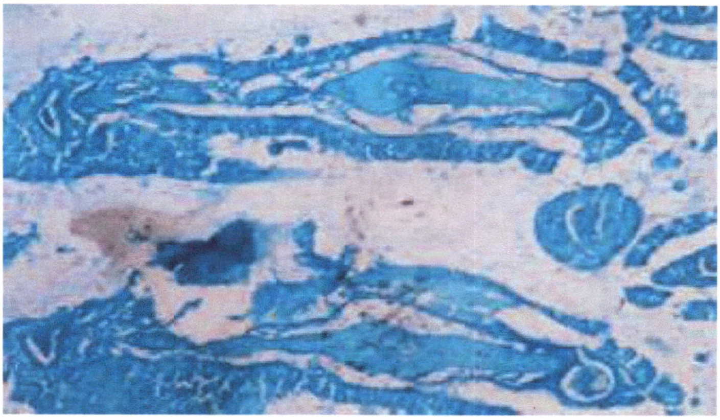

Fig. (6). T.S of *Channa striatus* infected gill ($KMnO_4$/Alician Blue).

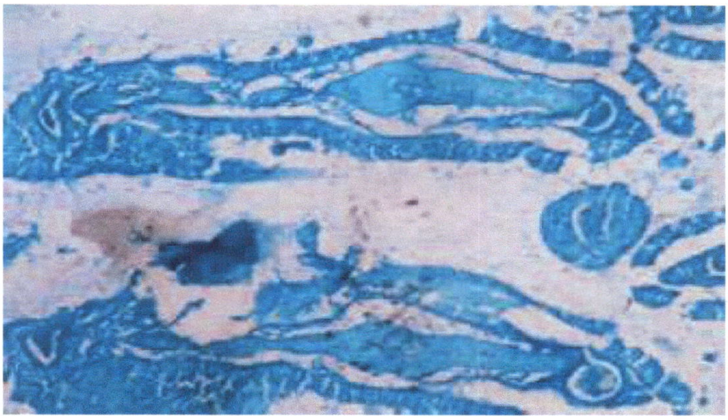

Fig. (7). T.S of *Channa punctatus* control gill ($KMnO_4$/Alician Blue).

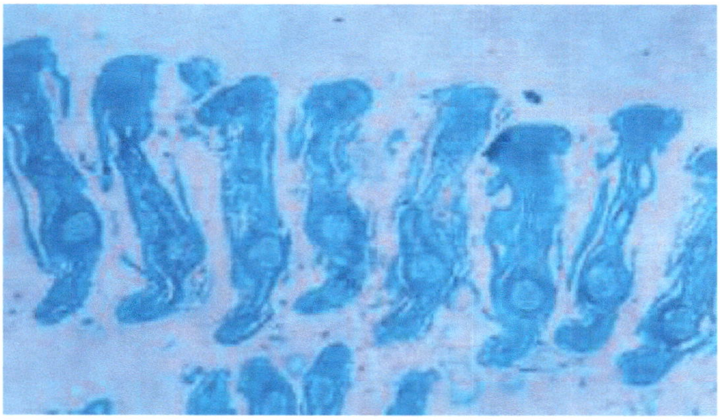

Fig. (8). T.S of *Channa punctatus* infected gill ($KMnO_4$/Alician Blue).

Copper Phthalocyanine (CPP)

The epithelial cells of the gill lamellae envelop the pillar cell system, collectively constituting the respiratory surface area of the gill lamellae. The gill epithelial mucous cells retain substantial quantities of proteins and phospholipids. The histochemical assays performed on the gill of *C. Striatus* and *C. Punctatus* demonstrated the existence of substantial quantities of phospholipids that exhibited high positivity in control gill tissue when stained with copper phthalocyanin. In the EUS-infected gill tissue, there is a modest positive reaction to copper phthalocyanine. Consequently, the presence of diverse protein types is observed in the mucous of gill epithelial cells, reflecting the dynamic activity of proteins (Plate **XVIII**, Figs. **1** to **4**).

PLATE - XVIII

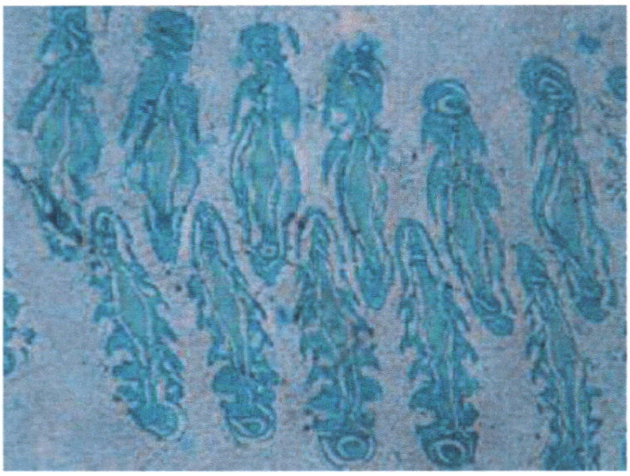

Fig. (1). T.S of *Channa striatus* control gill (CPP).

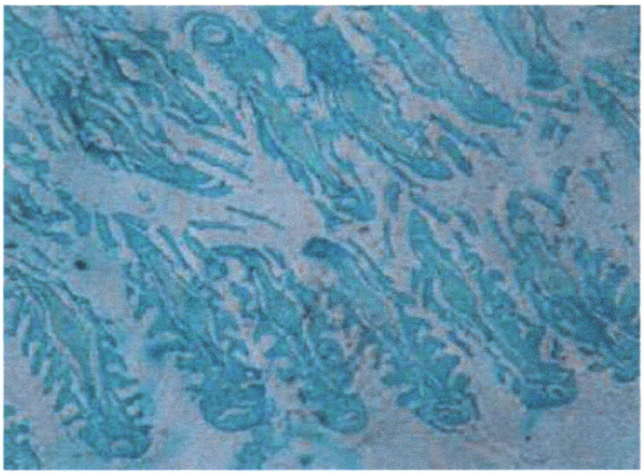

Fig. (2). T.S of *Channa striatus* infected gill (CPP).

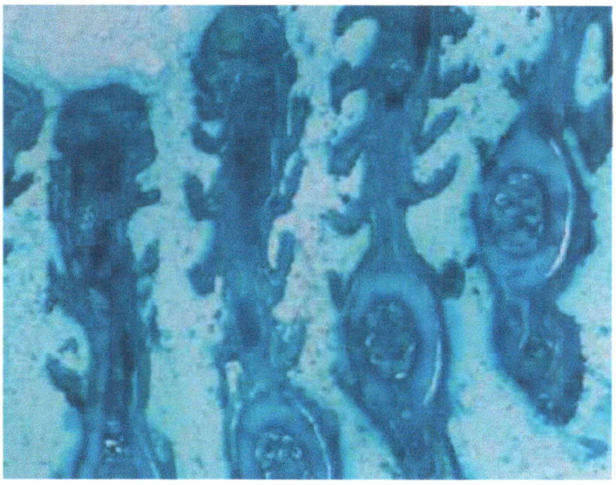

Fig. (3). T.S of *Channa punctatus* control gill (CPP).

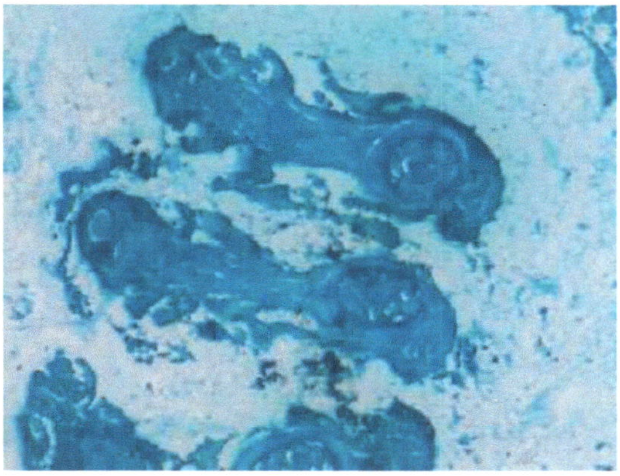

Fig. (4). T.S of *Channa punctatus* infected gill (CPP).

Nucleic Acids

Pyronin y

The Pyronin Y reaction in the control gill tissue produces a green/bluish-red coloration of variable intensities, reflecting the distribution of DNA and RNA inside the different cells of the primary and secondary lamellar epithelium. This test produces a green-blue color for DNA due to methyl green, whereas pyronin adds a red color upon reacting with RNA in *C. striatus* and *C. punctatus*.

The cytoplasm of mucous cells and their secretory products become more compacted and exhibit a pronounced red coloration, indicating a high RNA content. The nuclei of the supporting cells and receptor cells inside the gill epithelium exhibit DNA positivity. The central core exhibits a moderate response to this test.

In fish infected with bacteria and fungi, the detection and localization of DNA and RNA within the primary and secondary lamellae epithelium of *C. Striatus* and *C. punctatus* are seen. The DNA content in a certain cell remains consistent during its entire lifespan. The fluctuations are likely linked to physiological alterations in the various lamellar epithelial cells. The present inquiry examines the pronounced response of DNA within the nuclei of receptor cells, attributed to the condensation of chromatin material during their active state in the analyzed fish species. Detectable levels of DNA in the nuclei of flagellated supporting cells indicate their significant role in cellular activity and movement. The accumulation of a comparatively greater quantity of RNA in the cytoplasm of mucous cells clearly indicates its significant function in glycoprotein synthesis, which is crucial for mucin formation (Plate **XIX**, Figs. **1** to **4**).

PLATE - XIX

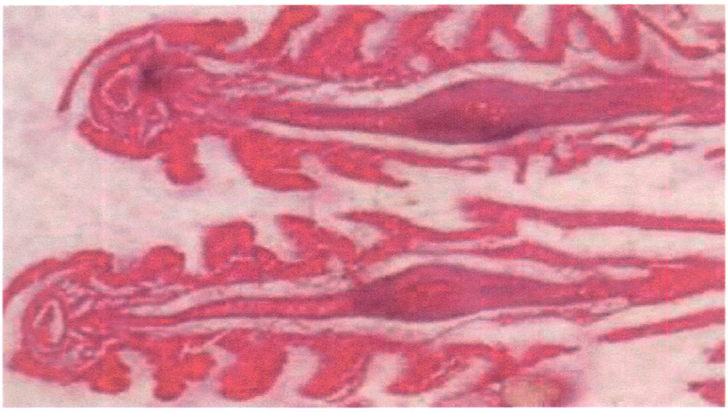

Fig. (1). T.S of *Channa striatus* control gill (Pyronin Y).

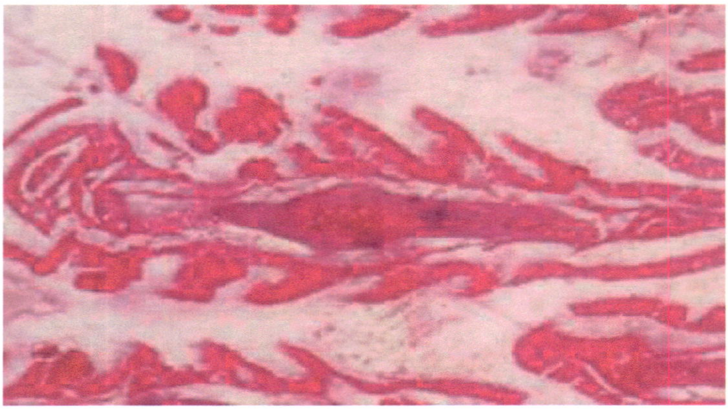

Fig. (2). T.S of *Channa striatus* infected gill (Pyronin Y).

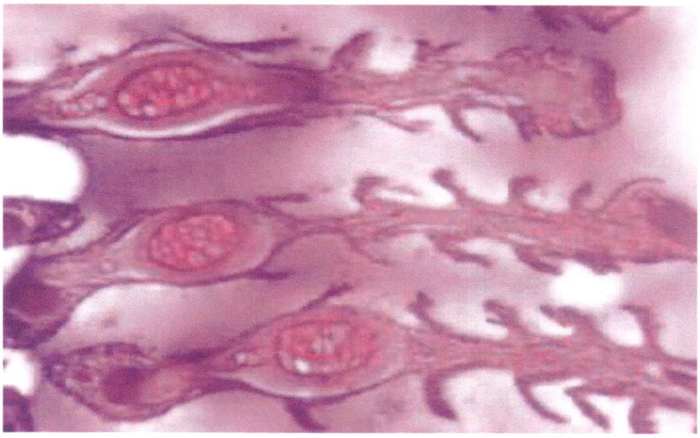

Fig. (3). T.S of *Channa punctatus* control gill (Pyronin Y).

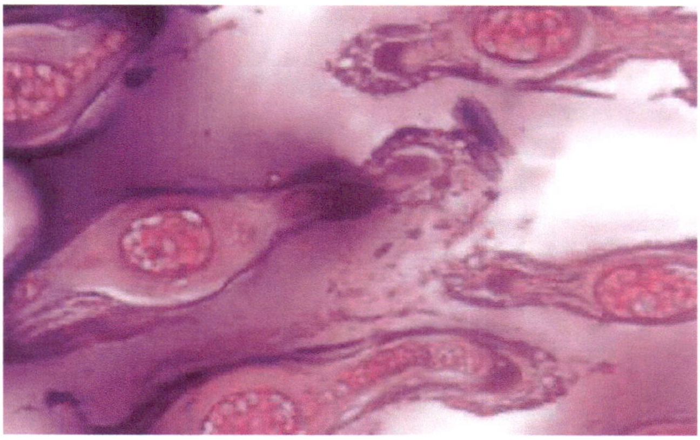

Fig. (4). T.S of *Channa punctatus* infected gill (Pyronin Y).

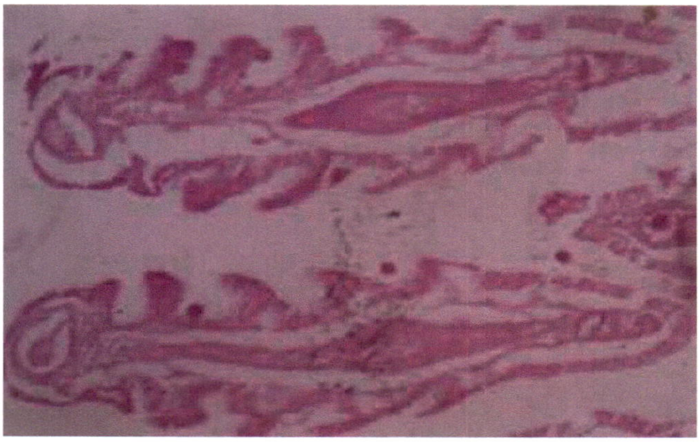

Fig. (5). T.S of *Channa striatus* control gill (Feulgen Reaction).

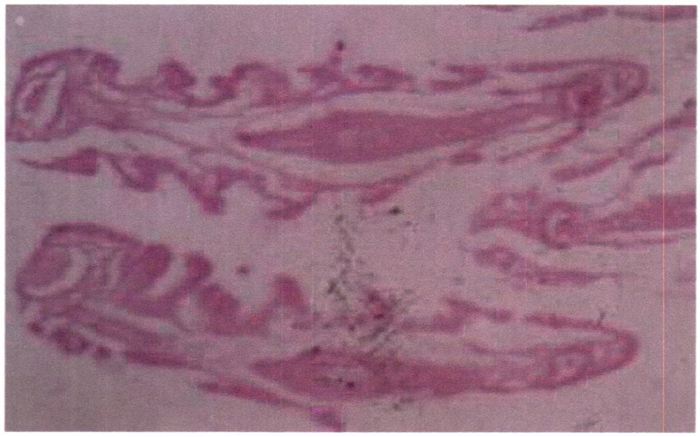

Fig. (6). T.S of *Channa striatus* infected gill (Feulgen Reaction).

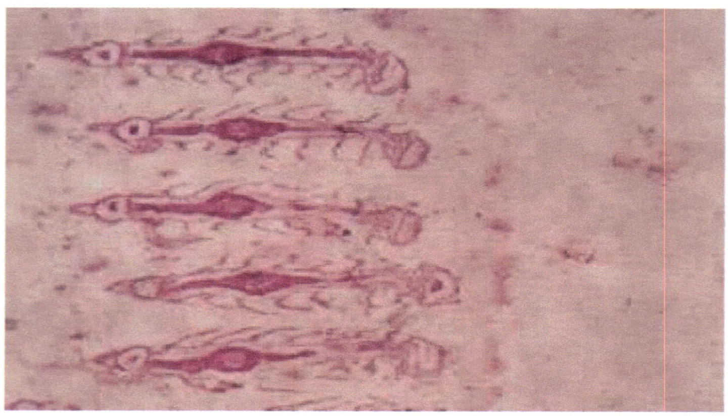

Fig. (7). T.S of *Channa punctatus* control gill (Feulgen Reaction).

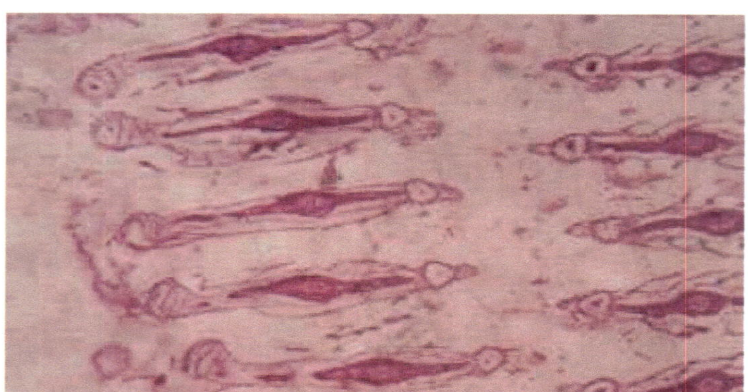

Fig. (8). T.S of *Channa punctatus* infected gill (Feulgen Reaction).

Feulgen Reaction

In *C. Striatus* and *C. punctatus*, the gill tissue has a robust response to the Feulgen reaction, indicating a significant presence of extracellular DNA in the control tissue sections.

Upon subjecting the bacterial and fungal-infected gill tissues to the Feulgen reaction, a moderate response was seen, indicating a depletion of DNA nucleic acids. This reveals an irregular and coarse lamellar epithelium aggregate irregularly distributed throughout the bronchial arches (Plate XIX, Figs. **5** to 8).

Comparative Histochemical Examination of Control and Diseased Liver

A series of histochemical approaches has been employed to clarify the chemical composition of normal and infected liver components. The pancreatic tissue comprises exocrine acinar cells and endocrine islet cells. Both cell types often exhibit analogous responses to all histochemical interactions.

Glycogen Concentration

Periodic Acid-Schiff (PAS) Reaction

In the control liver, hepatocytes exhibit a moderate reaction to the PAS approach, revealing a significant quantity of polysaccharide material in the cytoplasm as indicated by the PAS staining. A significant reduction of polysaccharides was noted in the cytoplasm of hepatocytes within the diseased liver.

The normal pancreatic tissue exhibits significant positivity with PAS. The tissue exhibited considerable staining following diastase digestion. The hue of the PAS remained intact upon acetylation. The pancreas tissue exhibited moderate positivity with Schiff's reagent, without preceding oxidation or periodic acid treatment. The infected pancreatic tissue had intensely discolored nuclei.

Alcian Blue (Ab) pH 2.5

Normal liver cells exposed to AB at pH 2.5. Weakly sulfated acid mucosubstances, hyaluronic acid, sialomucins, and sulfated mucosubstances are present. The infected liver exhibits an absence of mucin.

The pancreatic tissue exhibits significant positivity to AB at pH 2.5. This signifies the existence of weakly acidic sulfated mucosubstances, hyaluronic acid, sialomucins, and sulfated mucosubstances, but the infected pancreatic tissue exhibited a lack of mucin.

Alcian Blue (Ab) pH 1.0

At pH 1.0, the control hepatocytes exhibited red staining, signifying the presence of sialomucins. The liver cells infected by bacteria and fungi exhibited a blue-purple stain, signifying the presence of weakly acidic sulfated mucosubstances.

The pancreatic tissue exhibits significant positivity to AB pH 1.0, indicating the presence of acidic sulfated mucosubstances, hyaluronic acid, sialomucins, and sulfated mucosubstances. The infected pancreatic tissue exhibited mild staining, indicating the presence of insufficiently acidic sulfated mucosubstances (Plate **XX**, Figs. **1** to **12**).

PLATE – XX

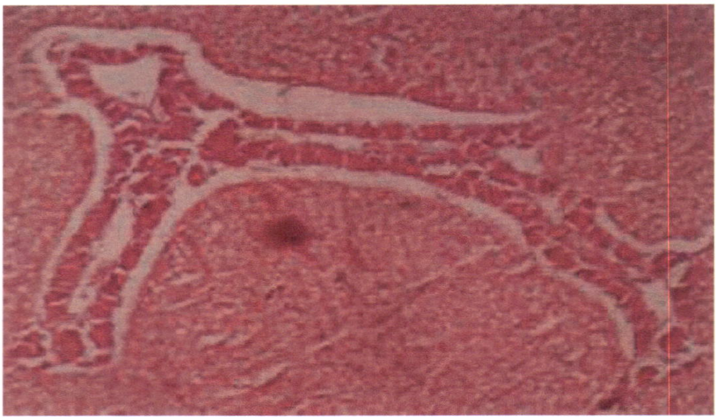

Fig. (1). T.S of control *Channa striatus* liver tissue displaying polysaccharides (PAS).

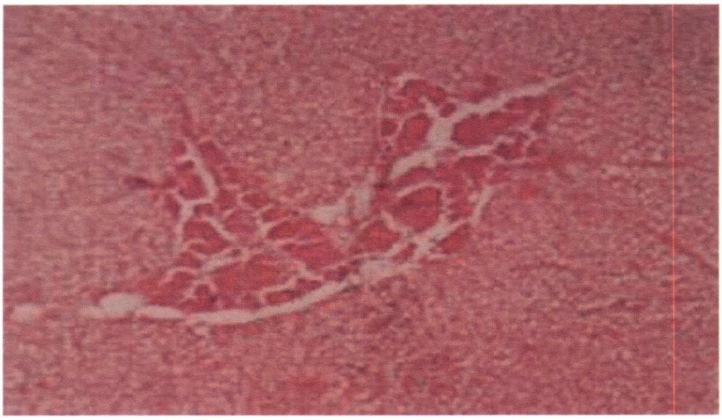

Fig. (2). T.S of *Channa striatus* infected liver tissue displaying absence of Glycogen (PAS).

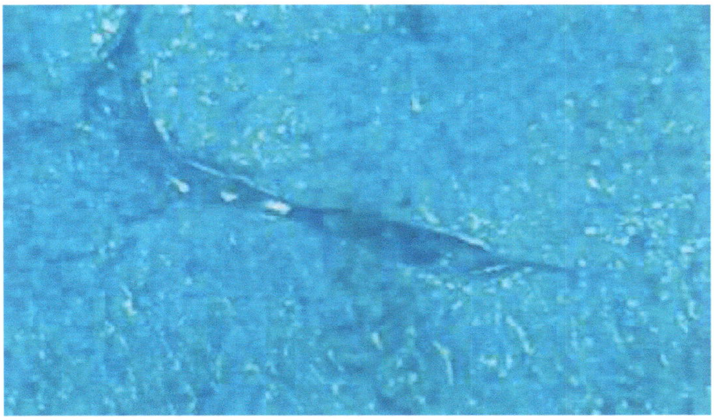

Fig. (3). T.S of *Channa striatus* control liver tissue displaying the mucosubstances (AB2.5pH).

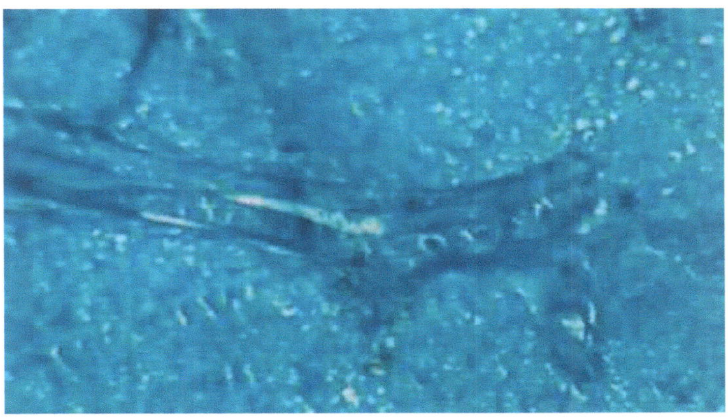

Fig. (4). T.S of infected *Channa striatus* liver tissue displaying absence of mucin (AB2.5pH).

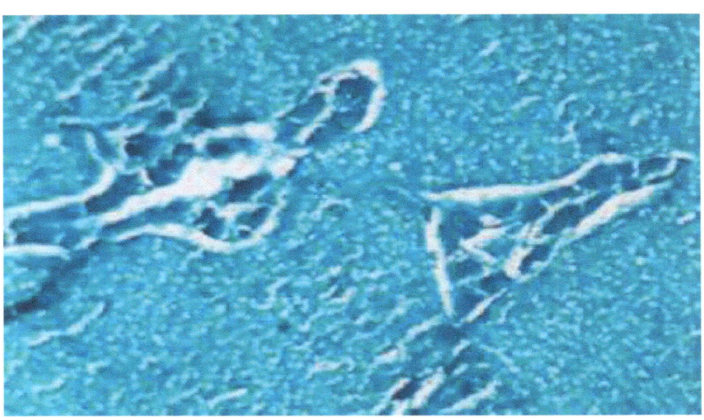

Fig. (5). T.S of *Channa striatus* control liver tissue exhibiting sialomucins (AB 1.0 pH).

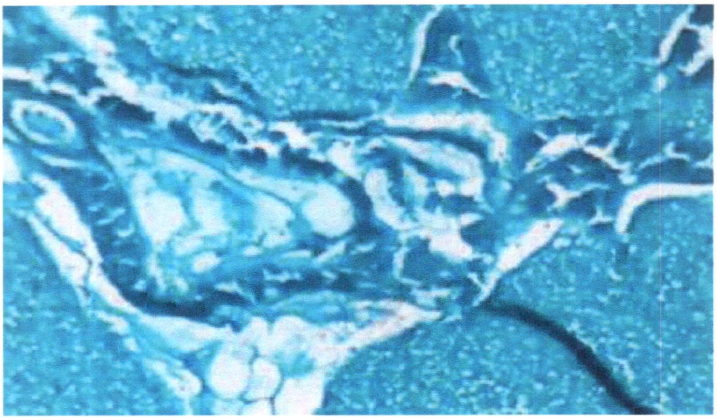

Fig. (6). T.S of *Channa striatus* infected liver tissue shows inadequately sulfated mucosubstances (AB 1.0 pH).

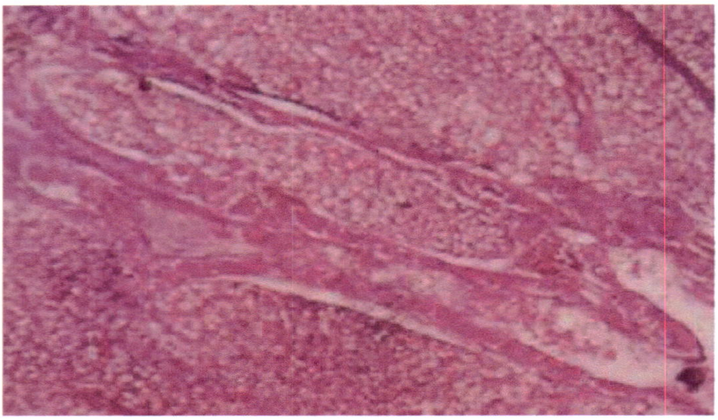

Fig. (7). T.S of *Channa punctatus* control liver tissue displaying polysaccharides (PAS).

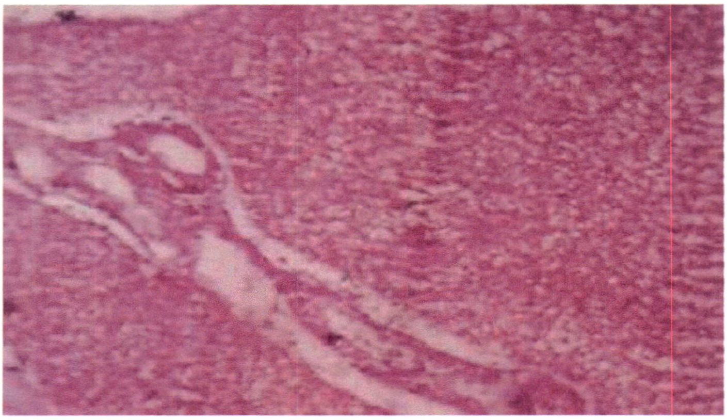

Fig. (8). T.S of *Channa punctatus* infected liver tissue displaying absence of Glycogen (PAS).

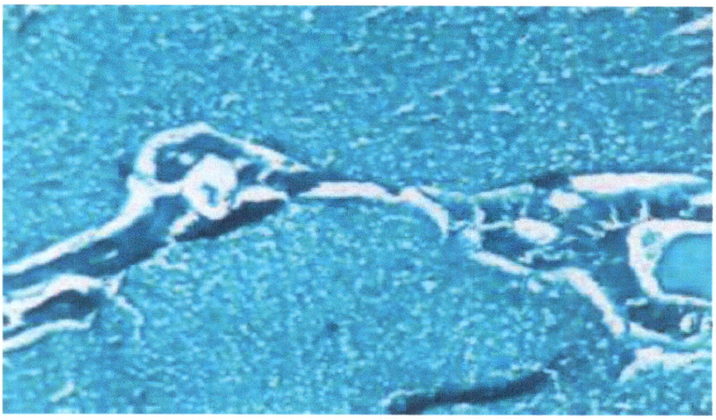

Fig. (9). T.S of *Channa punctatus* control liver tissue displaying the mucosubstances (AB2.5pH).

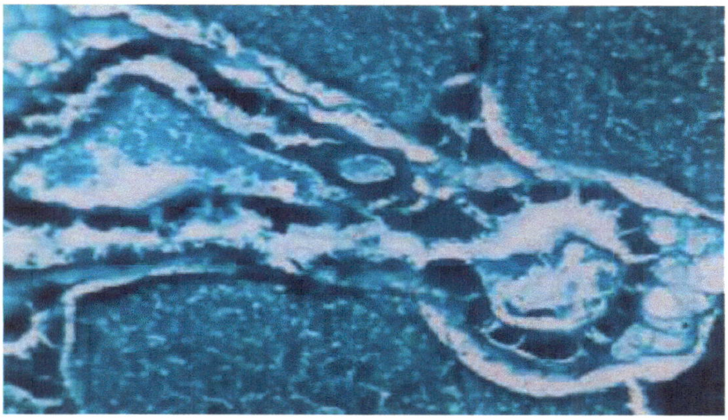

Fig. (10). T.S of *Channa punctatus* infected liver tissue displaying absence of mucin (AB2.5pH).

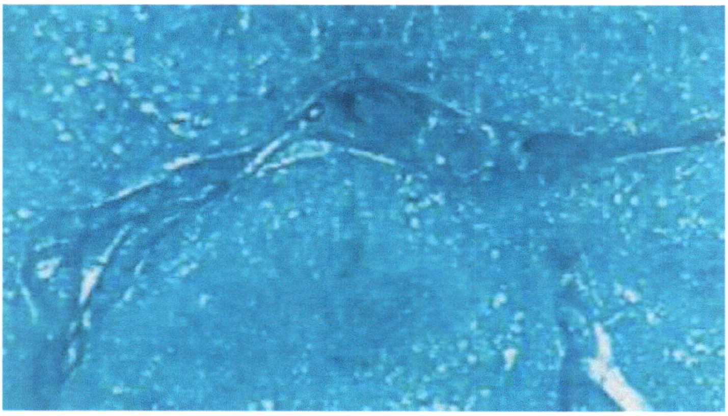

Fig. (11). T.S of *Channa punctatus* Control liver tissue exhibiting sialomucins (AB 1.0 pH).

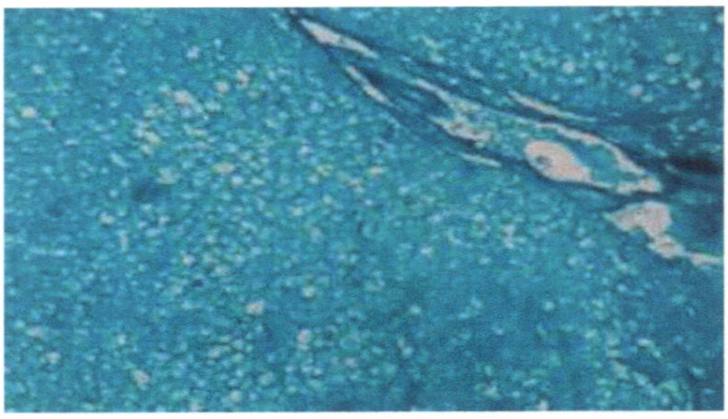

Fig. (12). T.S of *Channa punctatus* infected liver tissue shows inadequately sulfated mucosubstances (AB 1.0 pH).

Alcian Blue pH 2.5/PAS

In a normal liver, the majority of hepatocytes exhibit purple and blue coloration when subjected to the Alcian blue (pH 2.5)/PAS method, indicating the presence of sulfated mucosubstances. In the infected liver, the hepatocytes exhibited faint staining with AB (pH 2.5) in conjunction with PAS, attributable to the lack of both mucin and glycogen.

The normal pancreatic tissue exhibited a red coloration with AB pH 2.5/PAS, signifying the presence of sialomucins. Nevertheless, utilizing the combined method of AB pH 2.5 /PAS, the tissue of the diseased pancreas exhibited a slight blue stain, indicating the absence of mucins (Plate **XXI**, Figs. **1** to **4**).

Aldehyde Fuchsine

The aldehyde fuchsine staining reveals considerable positivity in the control liver cells, confirming the existence of collagen and elastin within these cells. The infected liver exhibited a purple-blue stain, indicating the presence of weakly acidic sulfated mucosubstances (Plate **XXI**, Figs. **5** to **8**).

PLATE – XXI

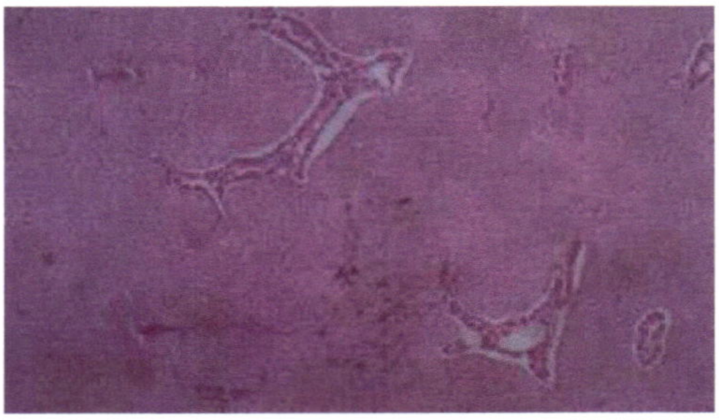

Fig. (1). T.S of *Channa striatus* control liver tissue displaying the presence of sulfated mucosubstances (AB 2.5 pH /PAS).

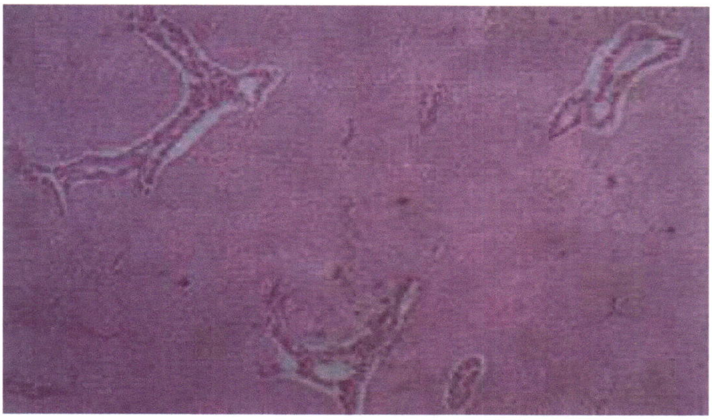

Fig. (2). T.S of *Channa striatus* infected liver tissue showing absence of both mucin and Glycogen (AB 2.5 pH/PAS).

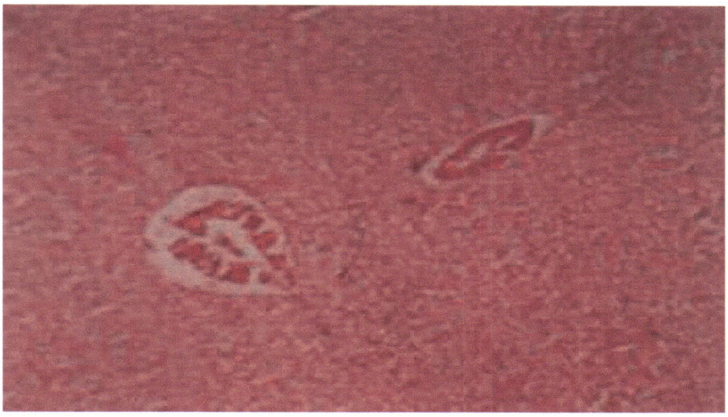

Fig. (3). T.S of *Channa striatus* control liver tissue exhibiting the mucopolysaccharides (Aldehyde fuchsine).

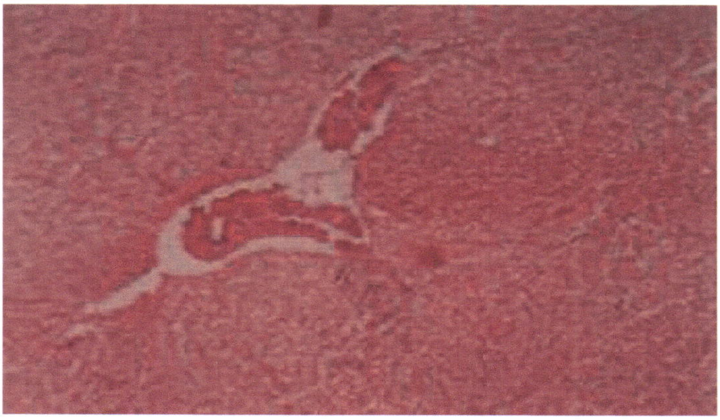

Fig. (4). T.S of *Channa striatus* infected silver tissue showing the weakly acidic sulfated mucosubstances (Aldehyde fuchsine).

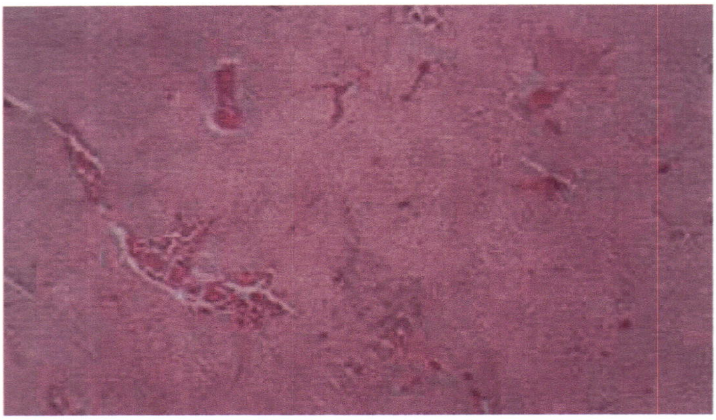

Fig. (5). T.S of *Channa punctatus* control liver tissue displaying the presence of sulfated mucosubstances (AB 2.5 pH /PAS).

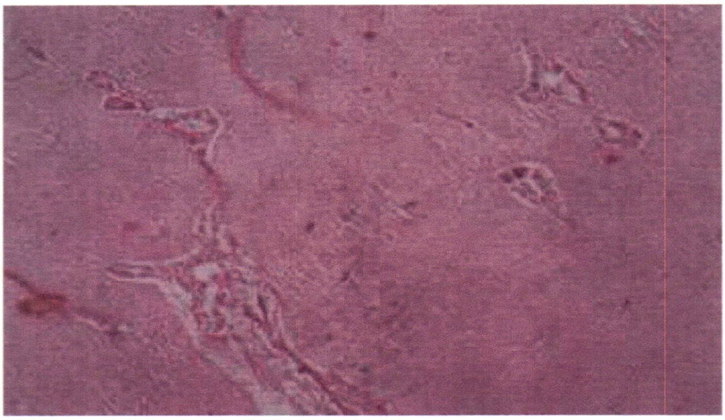

Fig. (6). T.S of *Channa punctatus* infected liver tissue showing absence of both mucin and Glycogen (AB 2.5 pH/PAS).

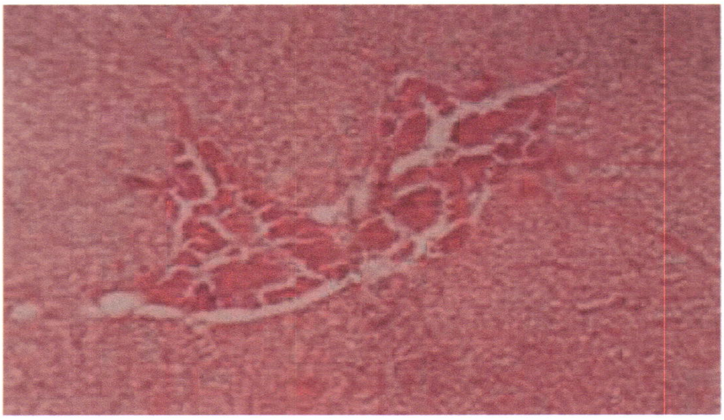

Fig. (7). T.S of *Channa punctatus* control liver tissue exhibiting the mucopolysaccharides (Aldehyde fuchsine).

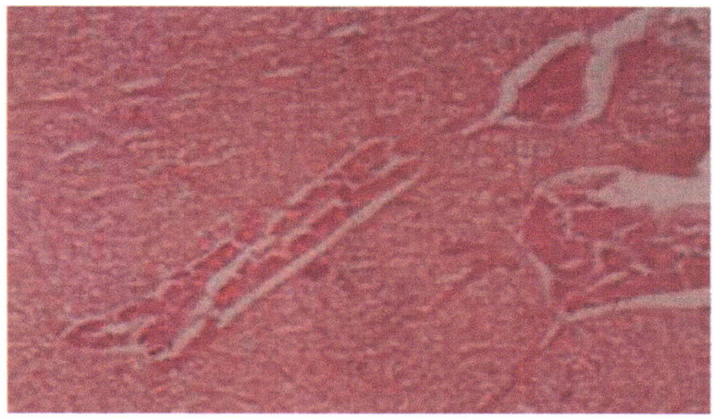

Fig. (8). .S of *Channa punctatus* infected liver tissue showing the weakly acidic sulfated mucosubstances (Aldehyde fuchsine).

Proteins

Mercuric Bromophenol Blue

The normal hepatocytes and pancreatic tissue exhibited strong positive staining for Bromophenol blue. This demonstrated the presence of both low and high concentrations of total protein in the tissues. The cytoplasm of hepatocytes exhibited a favorable reaction characterized by finely dispersed granulation.

The infected hepatocytes and pancreatic tissue exhibited moderate positivity. Total protein showed a significant reduction in the cytoplasm and nucleus of hepatocytes and pancreatic tissue.

The histochemical methods employed in this work have unequivocally demonstrated that in the two fish species. The liver includes a substantial quantity of proteinaceous material. This was confirmed when liver slices were submitted to additional specialized stains for the detection of various proteins (Plate **XXII**, Figs. **1** to **4**).

PLATE – XXII

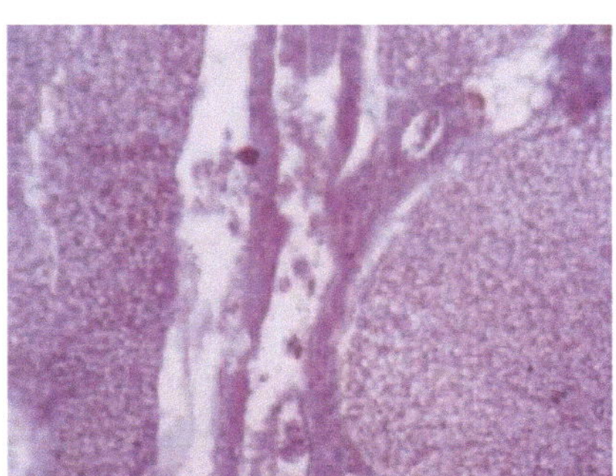

Fig. (1). T.S of *Channa striatus* control liver tissue displaying the total basic protein substances (Bromophenol blue).

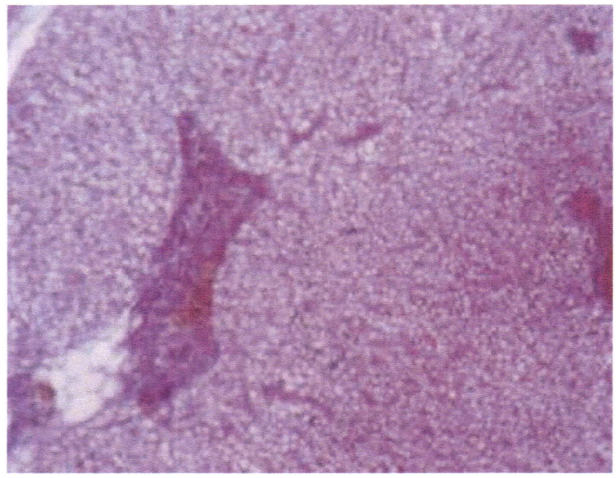

Fig. (2). T.S of *Channa striatus* infected liver tissue showing the basic proteins (Bromophenol blue).

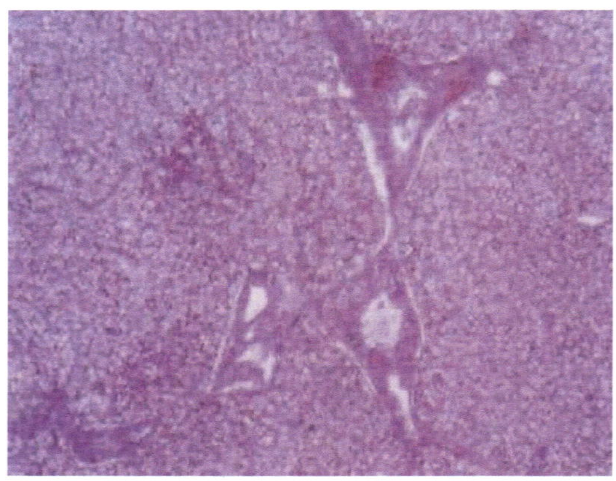

Fig. (3). T.S of *Channa punctatus* control liver tissue displaying the total basic protein substances (Bromophenol blue).

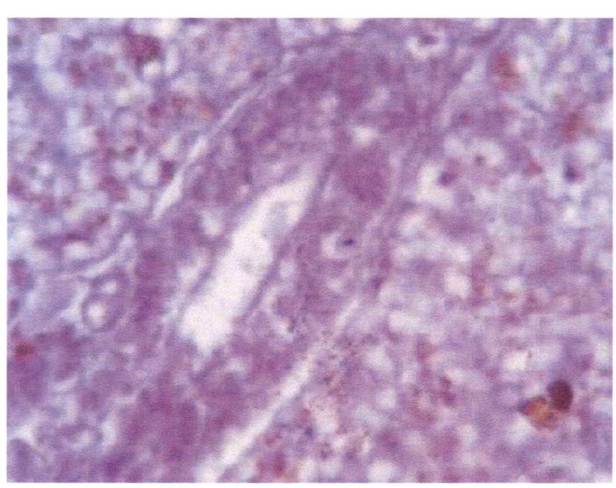

Fig. (4). T.S of *Channa punctatus* infected liver tissue showing the basic proteins (Bromophenol blue).

Congo Red

The normal hepatocytes and pancreatic cells exhibit significant positivity to Congo red. This positive indicated the existence of protein-bound amino groups and glycoproteins within the cells.

Infected hepatocytes and pancreas treated with Congo red exhibited a moderately positive effect. It indicated a total absence of protein, with remains primarily situated at the cell peripheries, exhibiting cytoplasmic vacuolation (Plate **XXIII**, Figs. **1** to **4**).

PLATE – XXIII

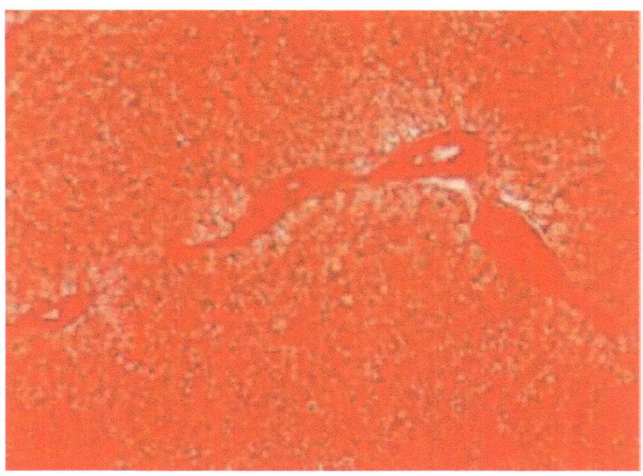

Fig. (1). T.S of *Channa striatus* control liver tissue showing protein-bound aminc groups and glycoproteins (Congo red).

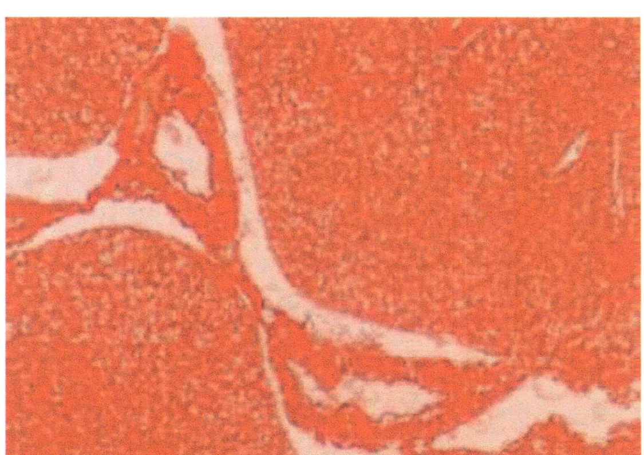

Fig. (2). T.S of *Channa striatus* infected liver tissue showing depletion of protein (Congo red).

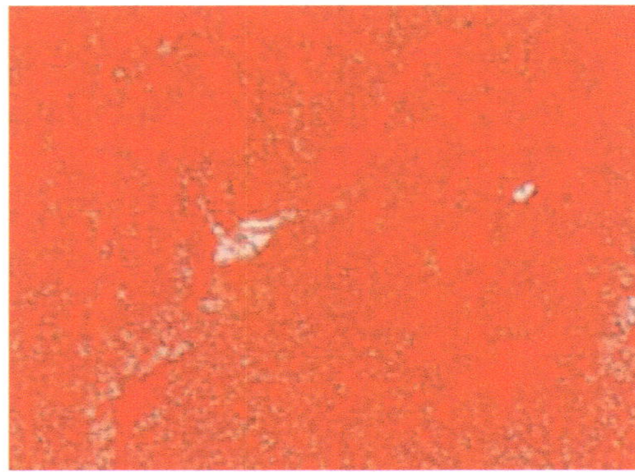

Fig. (3). T.S of *Channa punctatus* control liver tissue showing protein-bound amino groups and glycoproteins (Congo red).

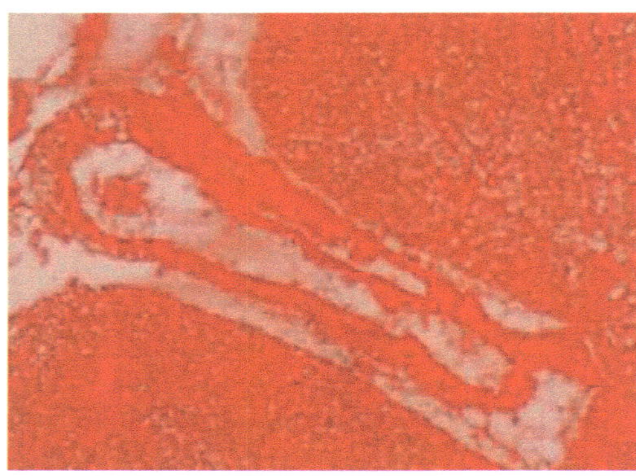

Fig. (4). T.S of *Channa punctatus* infected liver tissue showing depletion of protein (Congo red).

Lipid Composition

Sudan Black B

Sudan black B elicits a pronounced reaction in liver cells and pancreatic tissue, indicating the presence of lipids and phospholipids. Lipoid inclusion was evenly distributed within the cytoplasm, located perinuclearly and peripherally, with notable accumulation in the cytoplasm close to the sinusoid.

The infected liver and pancreas tissues, when exposed to Sudan black B, exhibited a moderate reaction, indicating a depletion of lipids and phospholipids. This resulted in an irregular and coarse aggregation of lipid material unevenly dispersed throughout the hepatic parenchyma. Conversely, the normal hepatopancreas displayed a regular distribution of lipids within the hepatic cells, evidenced by the presence of fine dark cytoplasmic granulation (Plate **XXIV**, Figs. **1** to **4**).

PLATE – XXIV

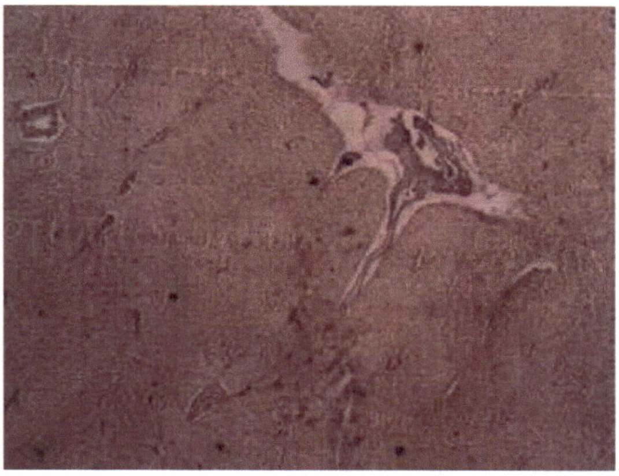

Fig. (1). T.S of *Channa striatus* control liver tissue exhibiting lipids and phospholipids (Sudan black B).

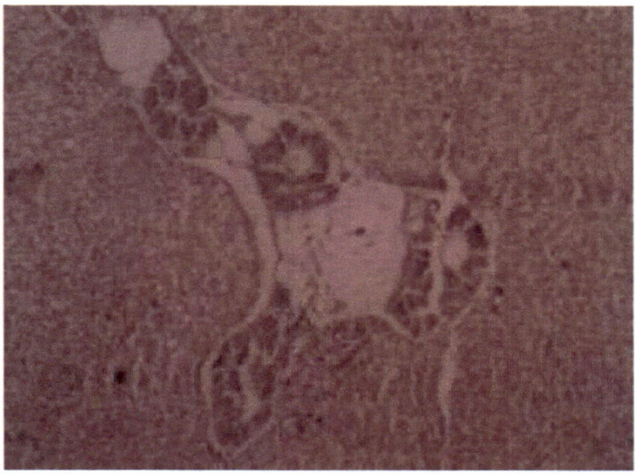

Fig. (2). T.S of *Channa striatus* infected liver tissue displaying depletion of lipids andphospholipids (Sudan black B).

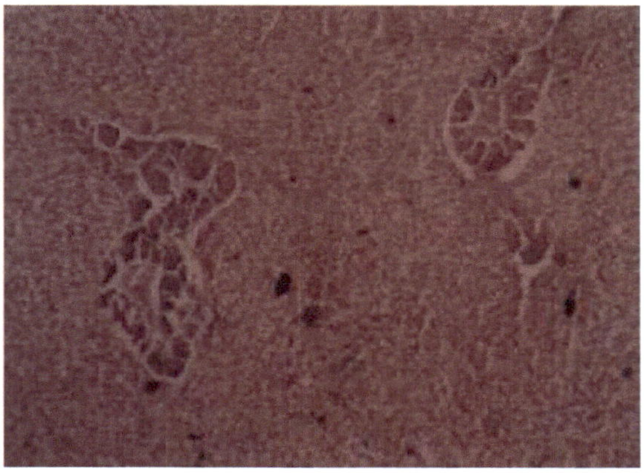

Fig. (3). T.S of *Channa punctatus* control liver tissue exhibiting lipids and phospholipids (Sudan black B).

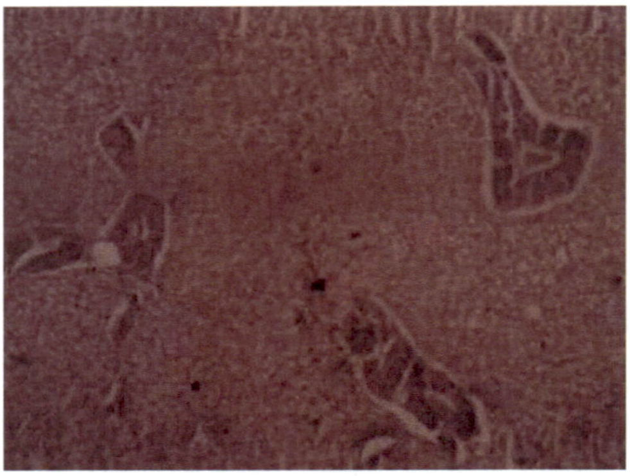

Fig. (4). T.S of *Channa punctatus* infected liver tissue displaying depletion of lipids and phospholipids (Sudan black B).

Ninhydrin/Schiff

The existence of protein-bound amino groups was confirmed using Ninhydrin/Schiff, yielding a moderately positive result in the normal liver tissue and a substantially positive result in the EUS-infected liver tissue (Plate **XXV**, Figs. **1** to **4**).

PLATE – XXV

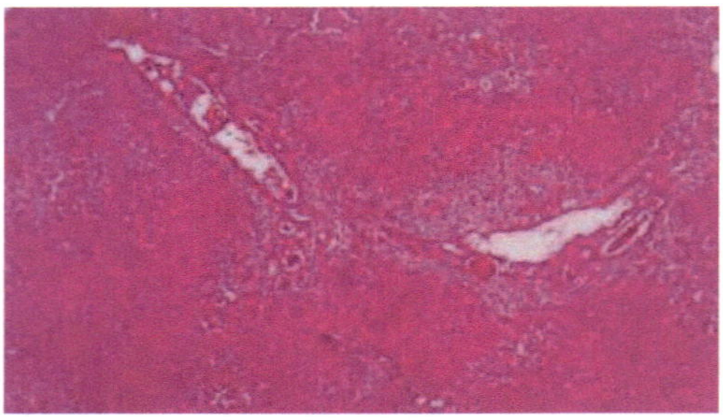

Fig. (1). T.S of *Channa striatus* control liver (Ninhydrin/Schiff).

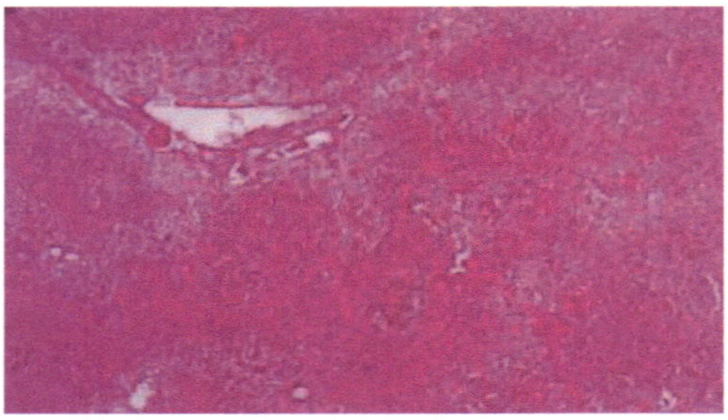

Fig. (2). T.S of *Channa striatus* infected liver (Ninhydrin/Schiff).

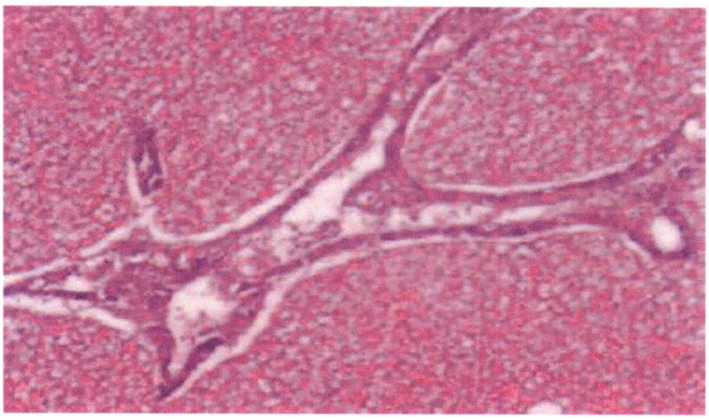

Fig. (3). T.S of *Channa punctatus* control liver (Ninhydrin/Schiff).

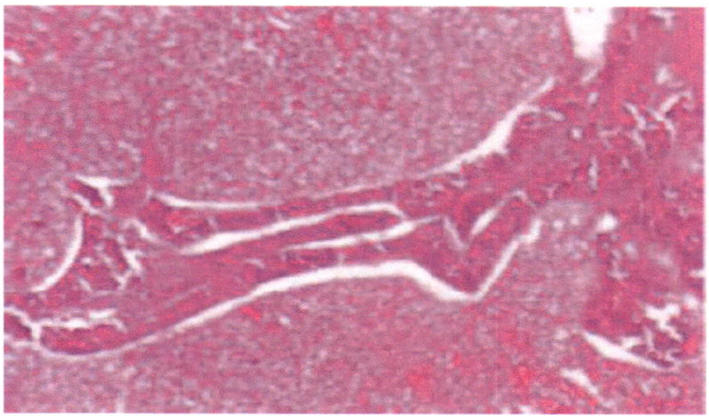

Fig. (4). T.S of *Channa punctatus* infected liver (Ninhydrin/Schiff).

Millon's Response

Amino acid tyrosine exhibited a somewhat favorable response in the control liver tissue and a positive response in the EUS-infected liver tissue, as indicated by Millon's reaction (Plate **XXV**, Figs. **5** to **8**).

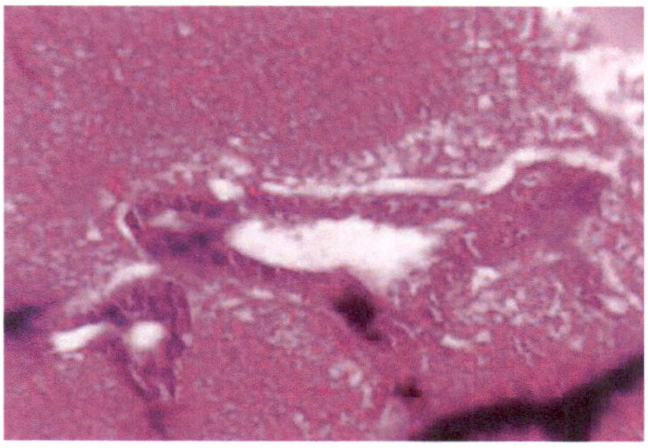

Fig. (5). T.S of *Channa striatus* control liver (Millon's reaction).

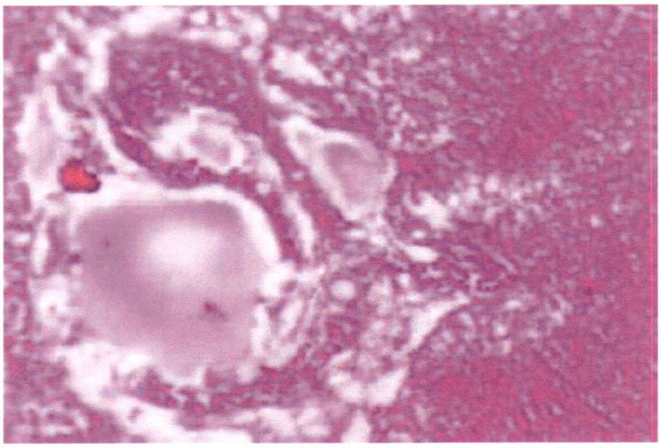

Fig. (6). T.S of *Channa striatus* Infected liver (Millon's reaction).

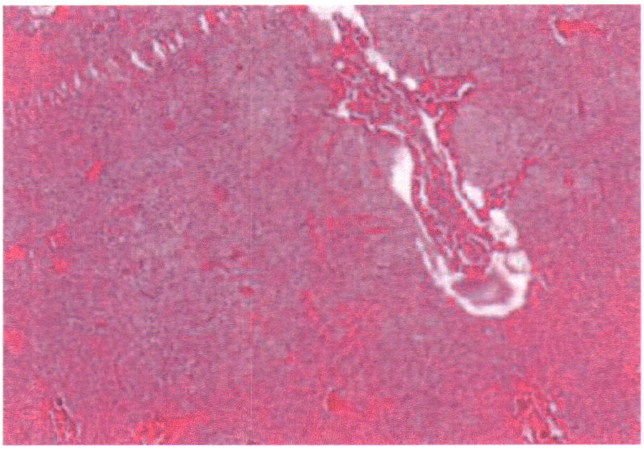

Fig. (7). T.S of *Channa punctatus* Control liver (Millon's reaction).

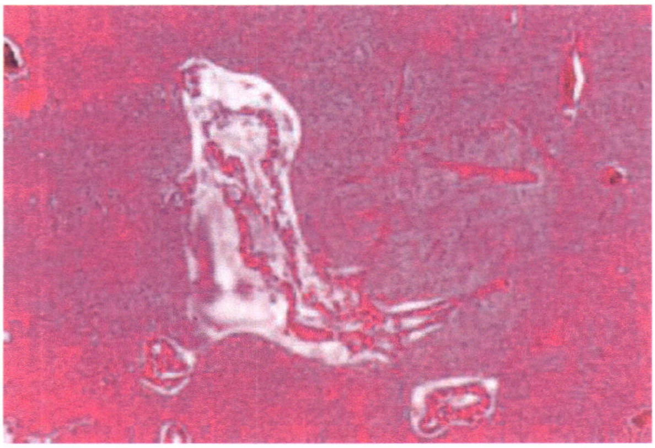

Fig. (8). T.S of *Channa punctatus* Infected liver (Millon's reaction).

Ferric Ferricyanide (ffc)

The control liver tissue exhibits modest levels of sulfhydryls, and disulfides were also identified. In contrast, the EUS-infected liver tissue demonstrated a favorable reaction to ferric ferricyanide (Plate **XXVI**, Figs. **1** to **4**).

PLATE - XXVI

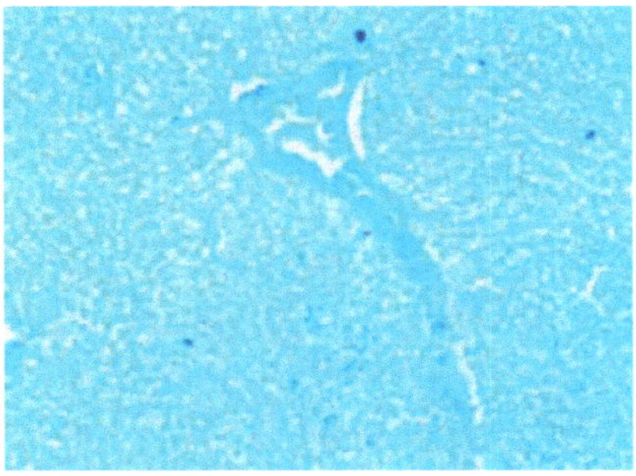

Fig. (1). T.S of *Channa striatus* control liver (FFC).

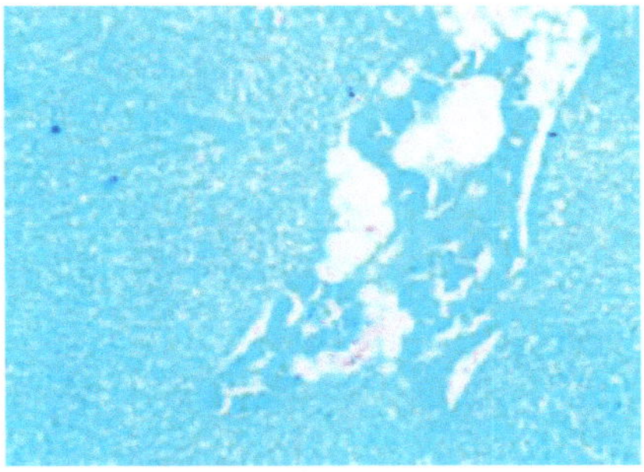

Fig. (2). T.S of *Channa striatus* infected liver (FFC).

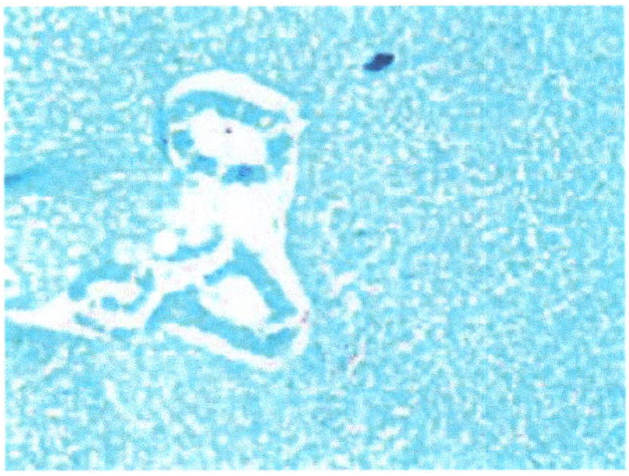

Fig. (3). T.S of *Channa punctatus* control liver (FFC).

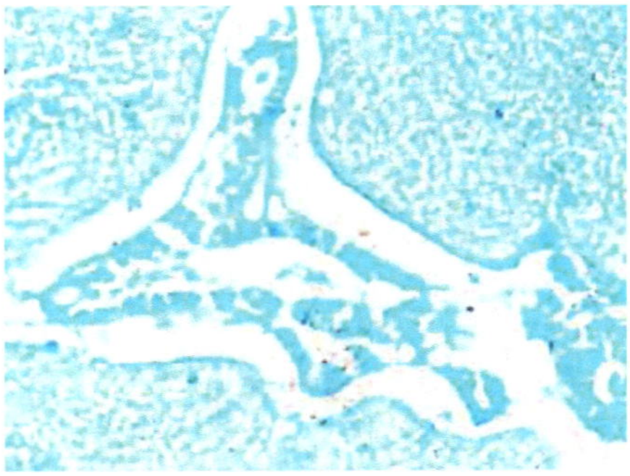

Fig. (4). T.S of *Channa punctatus* infected liver (FFC).

$KMnO_4$/AB

The control liver tissue exhibits a mild positive reaction to cystine with $Kmno_4$/Alcian blue, while the EUS-infected liver tissue demonstrates a positive response to $Kmno_4$/Alcian blue (Plate **XXVI**, Figs. **5** to **8**).

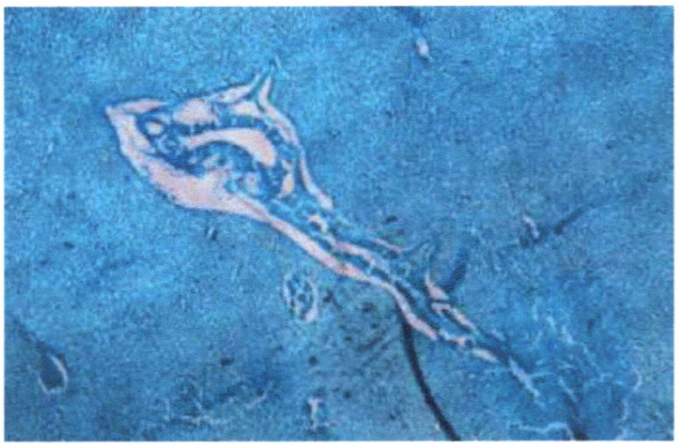

Fig. (5). T.S of *Channa striatus* control liver ($KMnO_4$/Alician Blue).

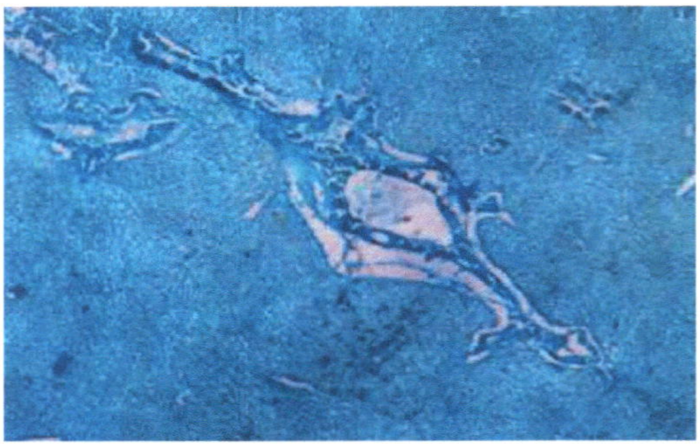

Fig. (6). T.S of *Channa striatus* infected liver ($KMnO_4$/Alician Blue).

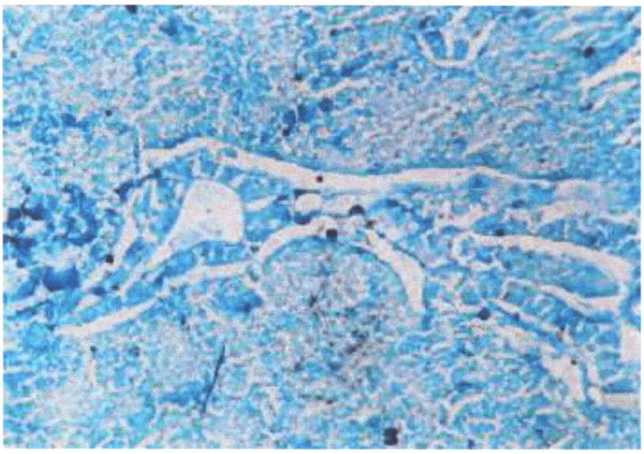

Fig. (7). T.S of *Channa punctatus* control liver ($KMnO_4$/Alician Blue).

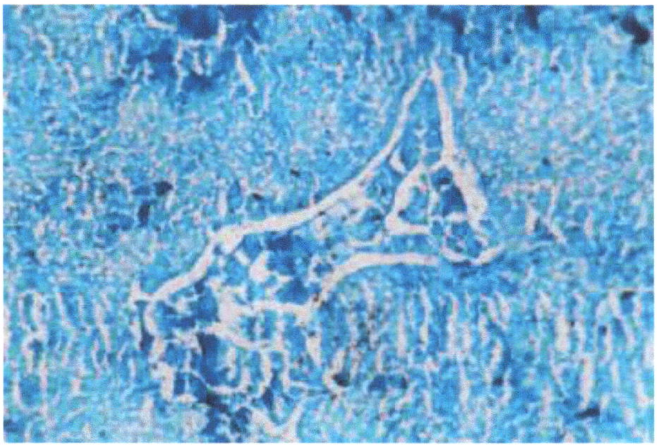

Fig. (8). T.S of *Channa punctatus* infected liver ($KMnO_4$/Alician Blue).

Copper PHTHALOCYANINE (CPP)

The liver functions as a reservoir for fats, facilitating their oxidation and energy release. It is also responsible for synthesizing fats from carbohydrates and proteins, as well as producing phospholipids. Histochemical analyses of hepatocytes from *C. striatus* and *C. punctatus* demonstrated substantial phospholipid presence, exhibiting strong positivity in control liver tissue when stained with copper phthalocyanine. In contrast, EUS-infected liver tissue showed a moderate positive response to copper phthalocyanine. Consequently, the liver is regarded as the locus for the specific dynamic action of proteins (Plate **XXVII**, Figs. **1** - **4**).

PLATE – XXVII

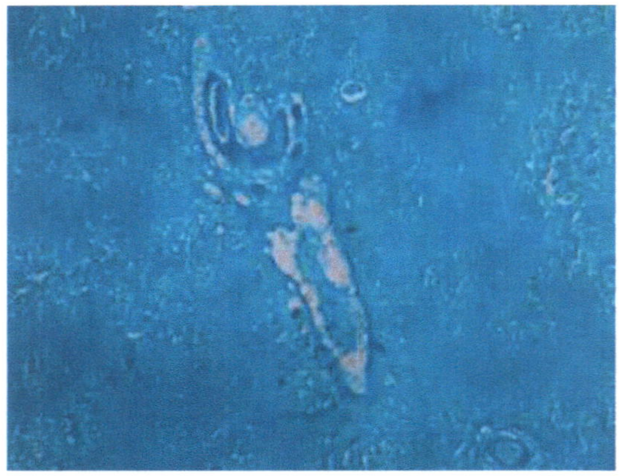

Fig. (1). T.S of *Channa striatus* control liver (CPP).

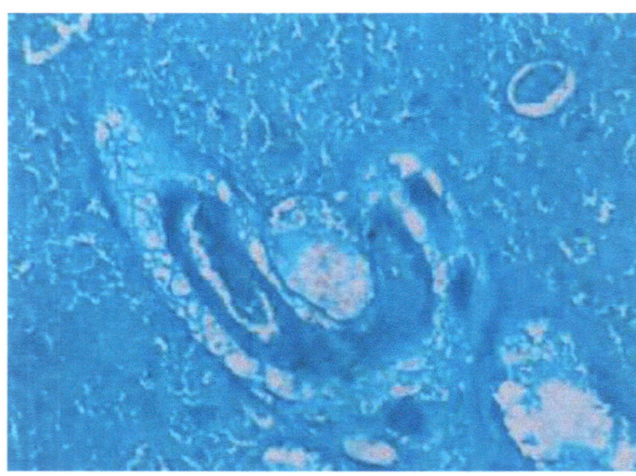

Fig. (2). T.S of *Channa striatus* infected liver (CPP).

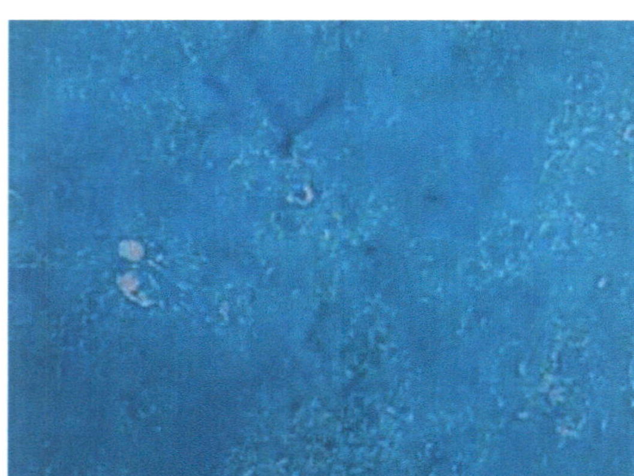

Fig. (3). T.S of *Channa punctatus* control liver (CPP).

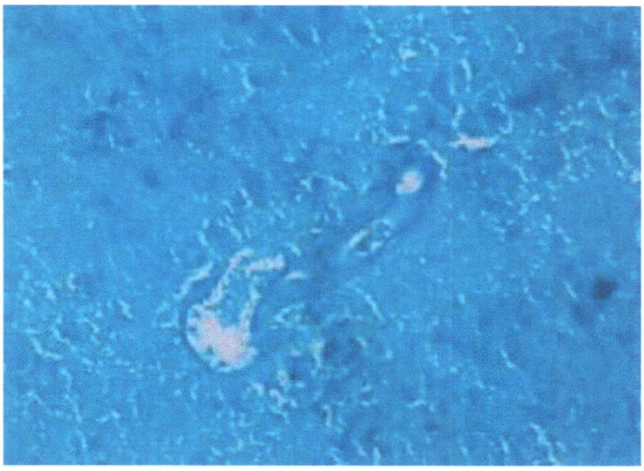

Fig. (4). T.S of *Channa punctatus* infected liver (CPP).

Nucleic Acids

Pyronin Y

The liver tissue has a high reaction to Pyronin Y, indicating the presence of extracellular RNA in the control tissue sections.

The liver tissues infected with bacteria and fungi exhibited a moderate reaction to Pyronin Y, indicating a depletion of RNA nucleic acids. This signifies a complete absence of nucleic acids, with remnants primarily located at the cell peripheries, particularly in the hepatic parenchymal cells (Plate **XXVIII**, Figs. **1** to **4**).

PLATE – XXVIII

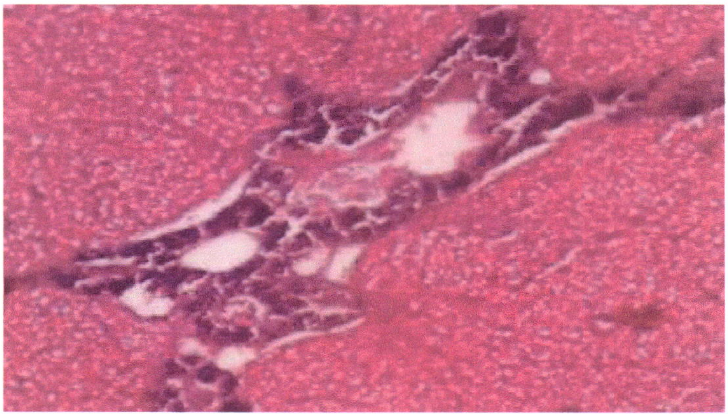

Fig. (1). T.S of *Channa striatus* control liver (Pyronin Y).

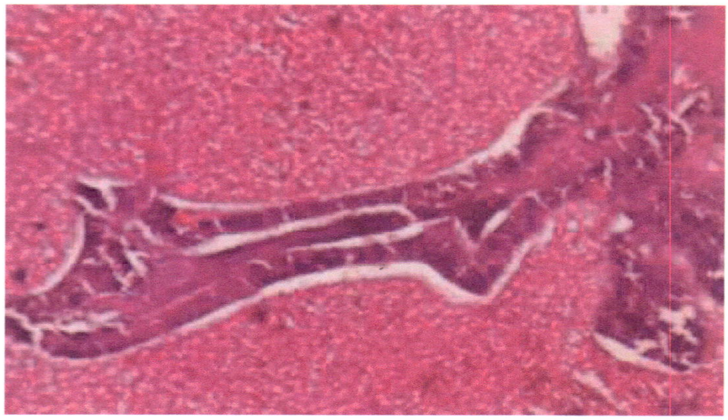

Fig. (2). T.S of *Channa striatus* infected liver (Pyronin Y).

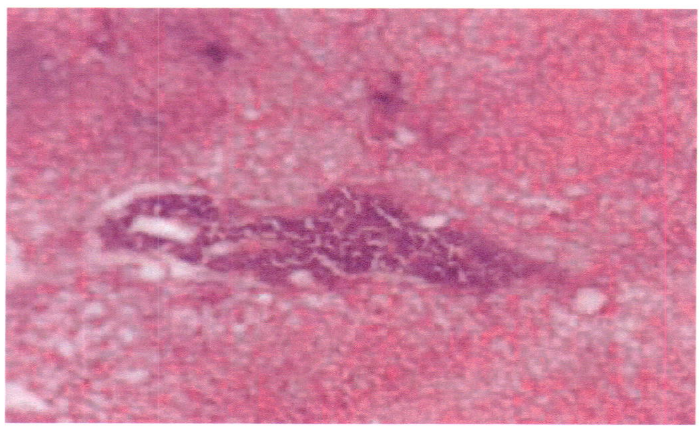

Fig. (3). T.S of *Channa punctatus* control liver (Pyronin Y).

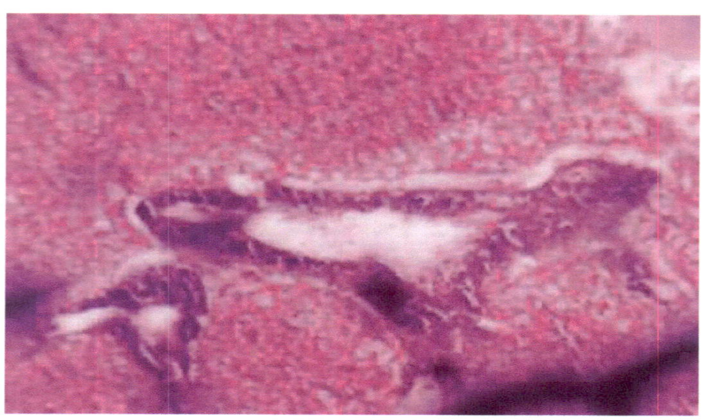

Fig. (4). T.S of *Channa punctatus* infected liver (Pyronin Y).

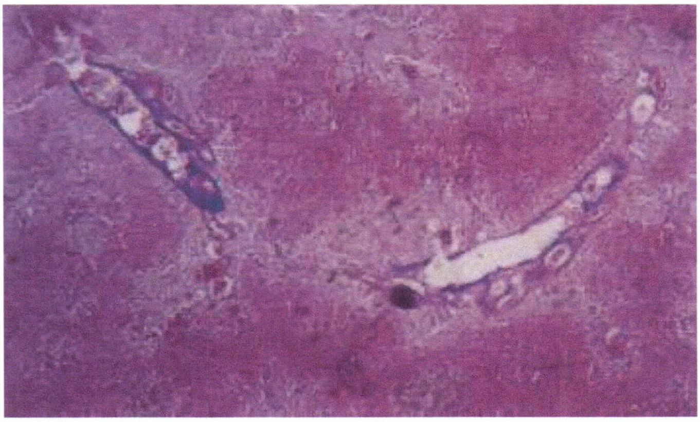

Fig. (5). T.S of *Channa striatus* control liver (Feulgen Reaction).

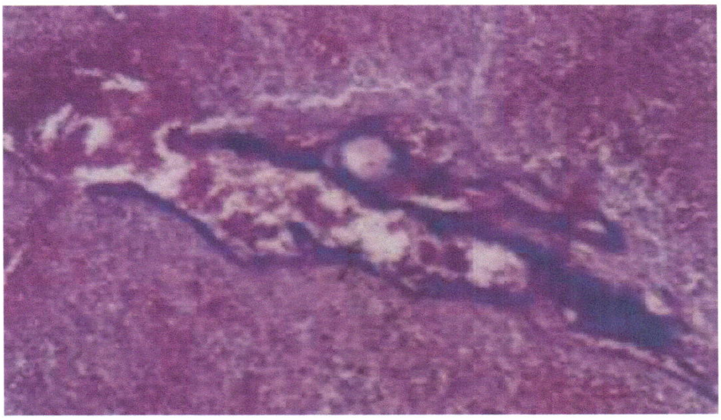

Fig. (6). T.S of *Channa striatus* infected liver (Feulgen Reaction).

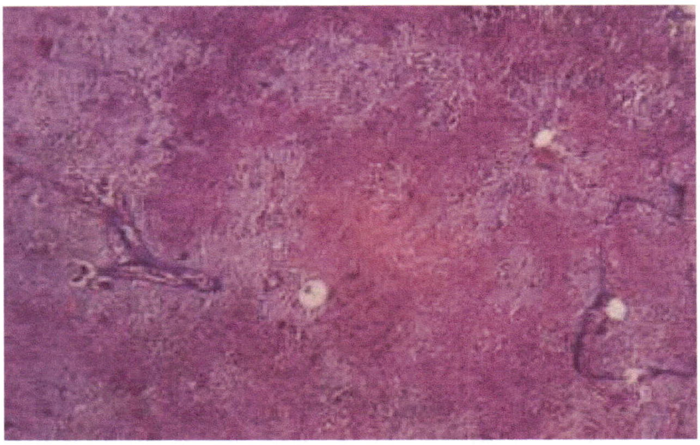

Fig. (7). T.S of *Channa punctatus* control liver (Feulgen Reaction).

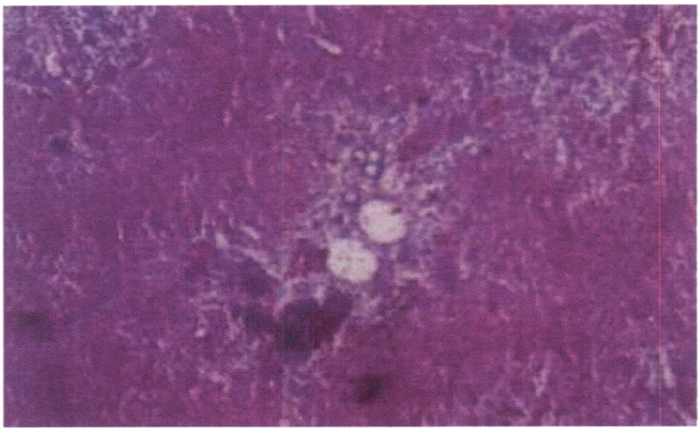

Fig. (8). T.S of *Channa punctatus* infected liver (Feulgen Reaction).

Feulgen Reaction

The control liver of fish exhibited a normal DNA content in the cytoplasmic hepatocytes. In contrast, the infected fish with bacterial and fungal infections displayed a moderate increase in liver DNA content compared to the control fish. Furthermore, there was no significant difference in liver DNA content among fish subjected to varying levels of pathogenic contamination.

The histochemical reactions indicated that the normal liver and pancreatic tissue of this fish comprised glycogen, weakly acidic sulfated mucosubstances, hyaluronic acid, sialomucins, carboxylated mucosubstances, basic and acidic proteins, sulfide and sulfhydryl groups of proteins, protein-bound amino groups, glycoproteins, lipids, phospholipids, elastin, and collagen.

The histochemical tests indicated that the infected liver and pancreatic tissues exhibit a depletion of glycogen, weakly acidic sulfated mucosubstances, hyaluronic acid, sialomucins, sulfated and carboxylated mucosubstances, basic and acidic proteins, sulfhydryl and disulfide groups of proteins, protein-bound amino groups, lipids, phospholipids, collagen, and elastin. The findings in the infected tissues were somewhat variable compared to normal liver and pancreatic tissues (Plate **XXVIII**, Figa. **5** to **8**).

Comparative Histochemical Examination of Control and Diseased Pancreatic Tissue

A series of histochemical assays has been utilized to elucidate the chemical composition of the pancreas. The Periodic Acid/Schiff reaction demonstrated a significant presence of carbohydrates in the pancreatic sections. The PAS/Saliva staining indicated moderate glycogen content. The elimination of PAS following acetylation and its restoration post-deacetylation confirmed the existence of 1:2

glycol groups. Free aldehydes were detected using Schiff's reagent in conjunction with periodic acid oxidation.

Bromophenol blue stain displayed a robust response to elevated concentrations of basic protein content in the pancreas. Millon's reaction was utilized to ascertain tyrosine content. Tryptophan was identified using the p-DMAB-nitrate staining technique, while cysteine was detected through the $KMNO_4$/alcian blue method. SH-bound proteins yielded a positive response to the Ferric-ferricyanide reagent. NH2-bound proteins were identified using the ninhydrin/Schiff reagent, and glycoproteins were revealed in substantial quantities following Congo red staining.

The Pyronin Y and Feulgen reactions confirmed the presence of RNA and DNA. The lipids and phospholipids were identified using Sudan Black B and Copper Phthalocyanine, with both classes of compounds exhibiting a robust response to these stains.

Glycogen Concentration

Periodic Acid-Schiff (PAS) Reaction

Normal pancreatic cells or hepatocytes exhibit a moderate reaction to the PAS technique, revealing a significant presence of polysaccharide material in the cytoplasm. Conversely, in the infected pancreas, a notable depletion of polysaccharides is observed in the cytoplasm of hepatocytes.

The control pancreatic tissue exhibited strong positivity with PAS. The tissue demonstrated moderate staining following diastase digestion. The PAS coloration was not diminished after acetylation. The pancreatic tissue displayed moderate positivity with Schiff's reagent, without prior oxidation or periodic acid treatment. The EUS-infected pancreatic tissue revealed intensely stained nuclei.

Alcian Blue (Ab) pH 2.5

The pancreatic tissue exhibits a high positive reaction to AB pH 2.5, indicating the presence of weakly acidic sulfated mucosubstances, hyaluronic acid, sialomucins, and sulfated mucosubstances. Conversely, the EUS-infected pancreatic tissue demonstrated an absence of mucin.

Alcian Blue (Ab) pH 1.0

At pH 1.0, normal hepatocytes exhibited red staining, signifying the presence of sialomucins. In contrast, the infected hepatocyte cells displayed blue-purple staining, suggesting the presence of weakly acidic sulfated mucosubstances.

The pancreatic tissue exhibits strong positivity to AB pH 1.0, indicating the presence of acidic sulfated mucosubstances, hyaluronic acid, sialomucins, and sulfated mucosubstances. In contrast, the infected pancreatic tissue displays faint staining, signifying the presence of inadequately acidic sulfated mucosubstances (Plate **XXIX**, Figs. **1** to **12**).

PLATE – XXIX

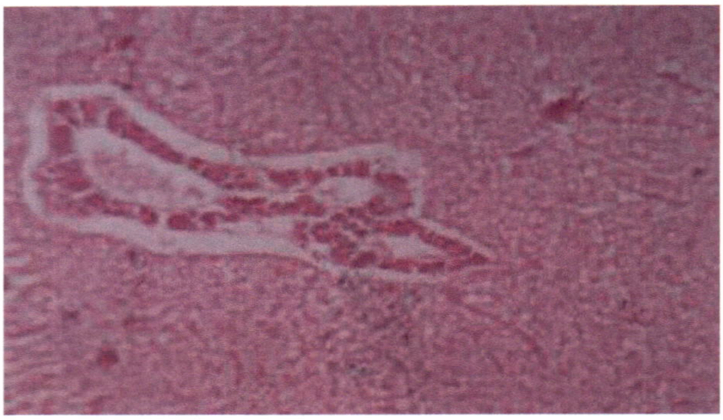

Fig. (1). T.S of *Channa striatus* control pancreas tissue displaying polysaccharides (PAS).

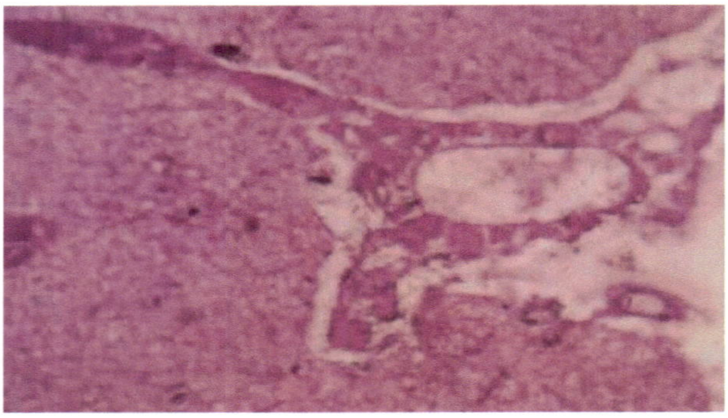

Fig. (2). T.S of *Channa striatus* infected pancreas tissue displaying absence of Glycogen (PAS).

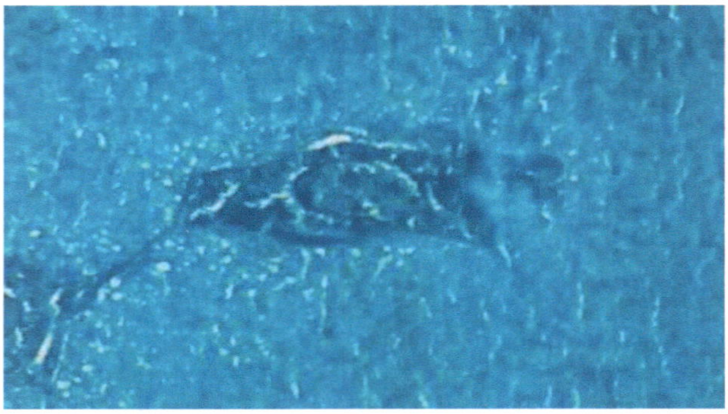

Fig. (3). T.S of *Channa striatus* control pancreas tissue displaying the mucosubstances (AB2.5pH).

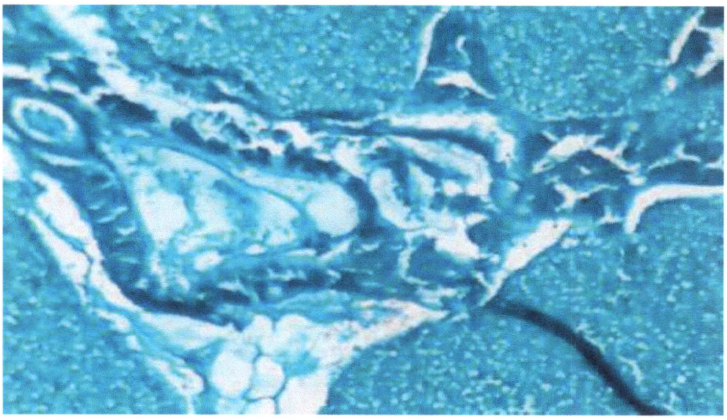

Fig. (4). T.S of *Channa striatus* infected pancreas tissue displaying absence of mucin (AB2.5pH).

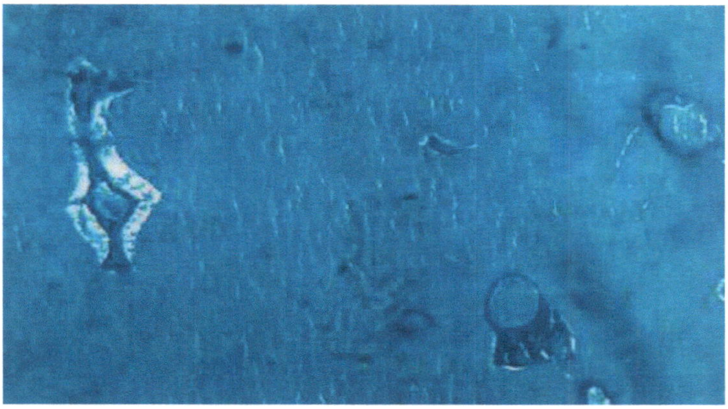

Fig. (5). T.S of *Channa striatus* control pancreas tissue exhibiting Sialomucins (AB 1.0 pH).

Fig. (6). T.S of *Channa striatus* infected pancreas tissue shows an inadequate sulfation of mucosubstances (AB 1.0 pH).

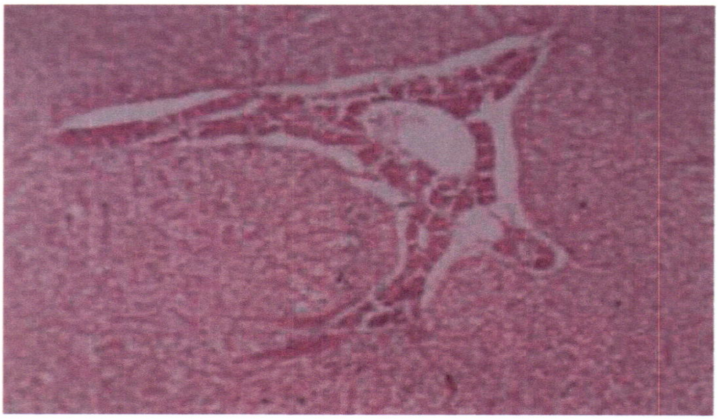

Fig. (7). T.S of *Channa punctatus* control pancreas tissue displaying polysaccharides (PAS).

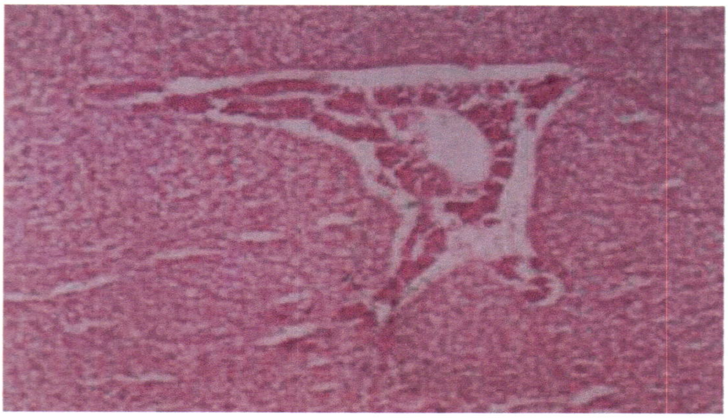

Fig. (8). T.S of *Channa punctatus* infected pancreas tissue displaying absence of Glycogen (PAS).

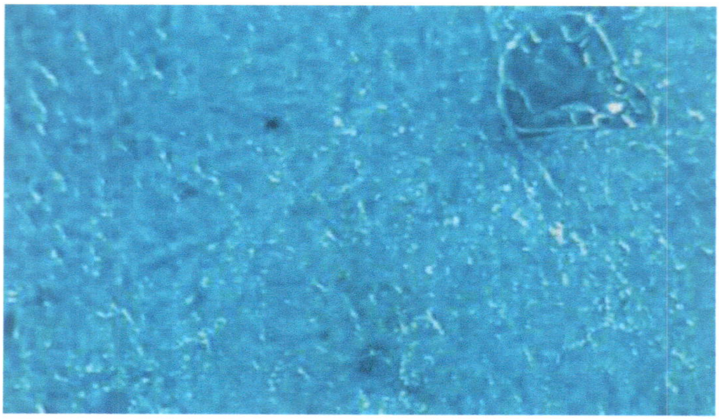

Fig. (9). T.S of *Channa punctatus* control pancreas tissue displaying the mucosubstances (AB2.5pH).

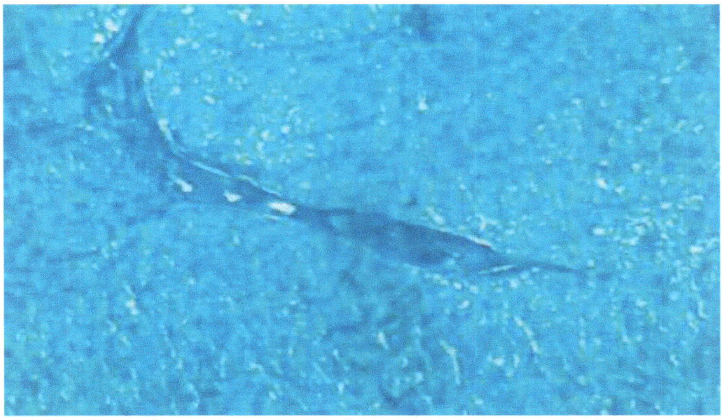

Fig. (10). T.S of *Channa punctatus* infected pancreas tissue displaying absence of mucin (AB2.5pH)

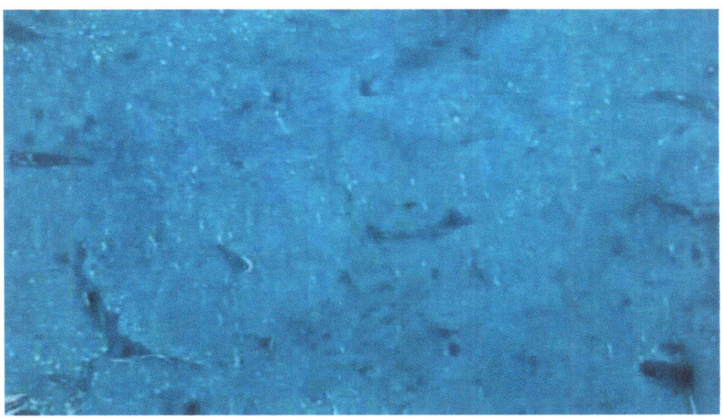

Fig. (11). T.S of *Channa punctatus* control pancreas tissue exhibiting sialomucins (AB 1.0 pH).

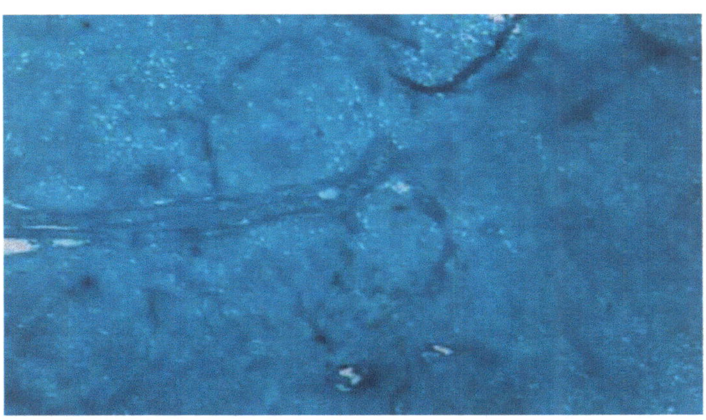

Fig. (12). T.S of *Channa punctatus* infected pancreas tissue shows inadequately sulfated mucosubstances (AB 1.0 pH).

Alcian Blue (Ab) pH 2.5/PAS

The normal pancreatic tissue exhibited a red coloration with AB pH 2.5/PAS, signifying the existence of sialomucins. Conversely, the pancreatic tissue affected by bacterial and fungal infections stained pale blue with the combined AB pH 2.5/PAS approach, suggesting a lack of mucins (Plate **XXX**, Figs. **1** to **4**).

Aldehyde Fuchsine

The normal pancreatic tissues exhibited significant positivity to Aldehyde fuchsin, indicating the presence of sulfated mucosubstances.

The pancreatic tissue infected by bacteria and fungi exhibited a blue-purple stain, signifying the presence of weakly acidic sulfated mucosubstances (Plate **XXX**, Figs. **5** to **8**).

PLATE – XXX

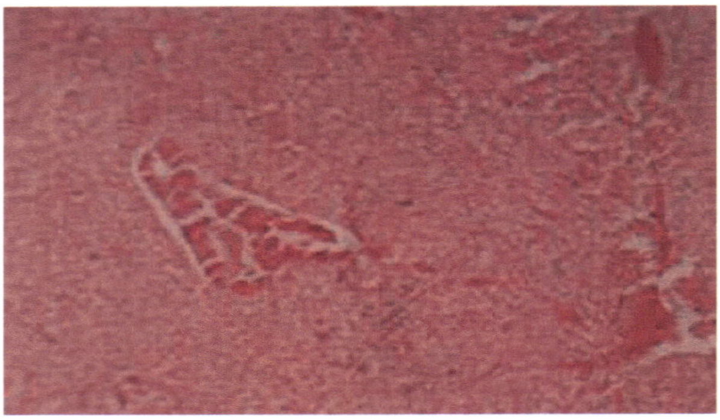

Fig. (1). T.S of *Channa striatus* control pancreas tissue displaying the presence of sulfated mucosubstances (AB 2.5 pH /PAS).

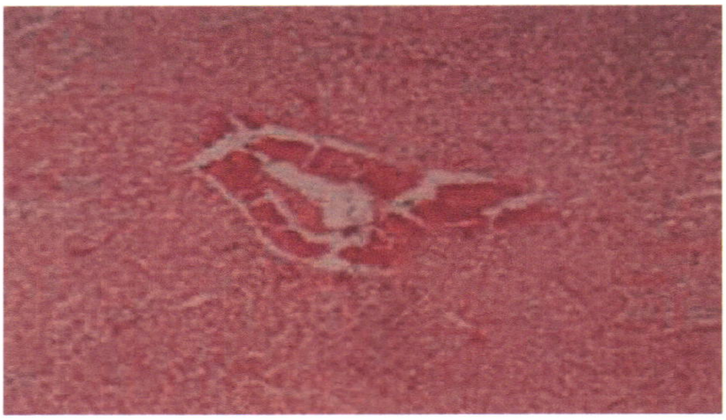

Fig. (2). T.S of *Channa striatus* infected pancreas tissue shows absence of both mucin and Glycogen (AB 2.5 pH/PAS).

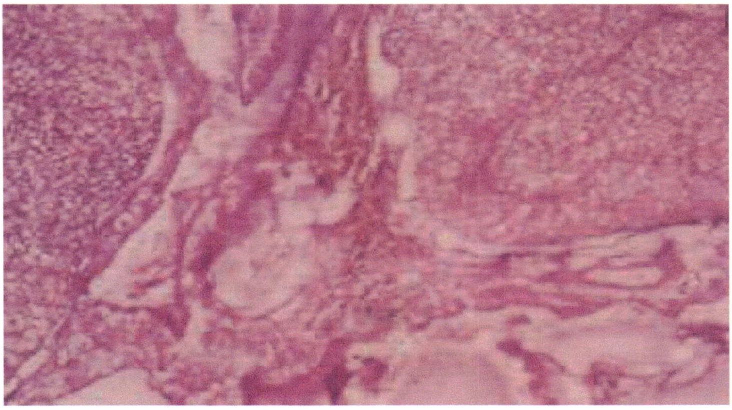

Fig. (3). T.S of *Channa striatus* control pancreas tissue exhibiting the mucopolysaccharides (Aldehyde fuchsine).

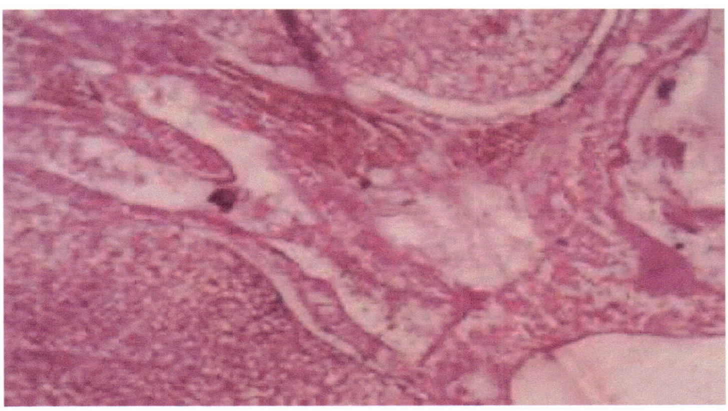

Fig. (4). T.S of *Channa striatus* infected pancreas tissue shows the weakly acidic sulfated mucosubstances (Aldehyde fuchsine).

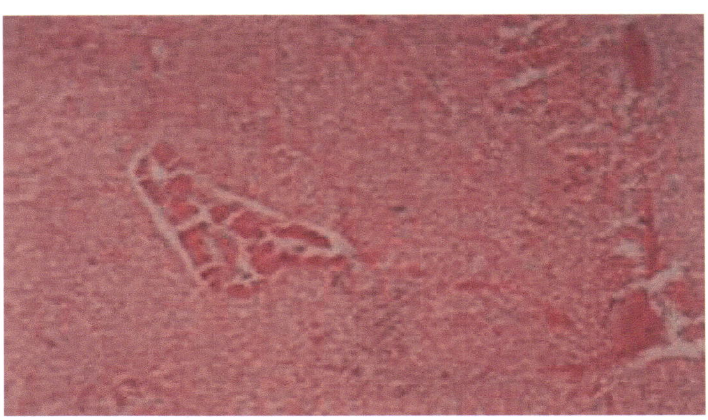

Fig. (5). T.S of *Channa punctatus* control pancreas tissue displaying the presence of sulfated mucosubstances (AB 2.5 pH /PAS).

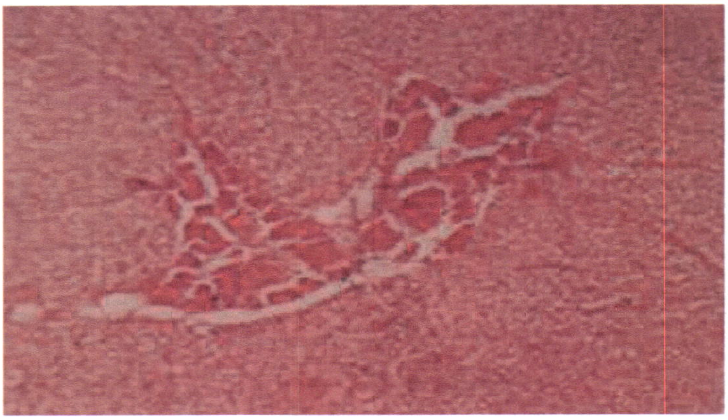

Fig. (6). T.S of *Channa punctatus* infected pancreas tissue shows absence of both mucin and glycogen (AB 2.5 pH/PAS).

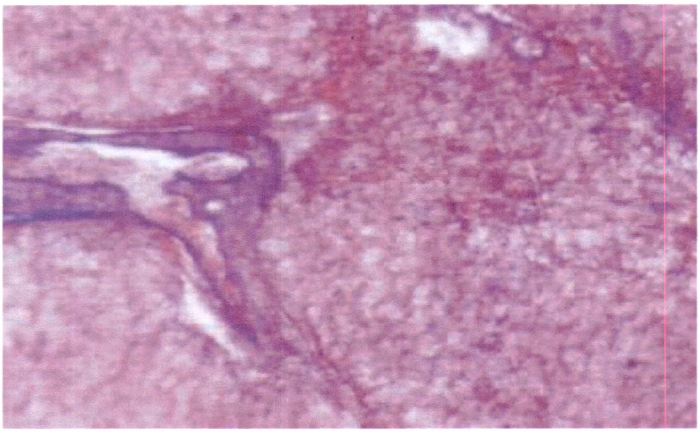

Fig. (7). T.S of *Channa punctatus* control pancreas tissue exhibiting the mucopolysaccharides (Aldehyde fuchsine).

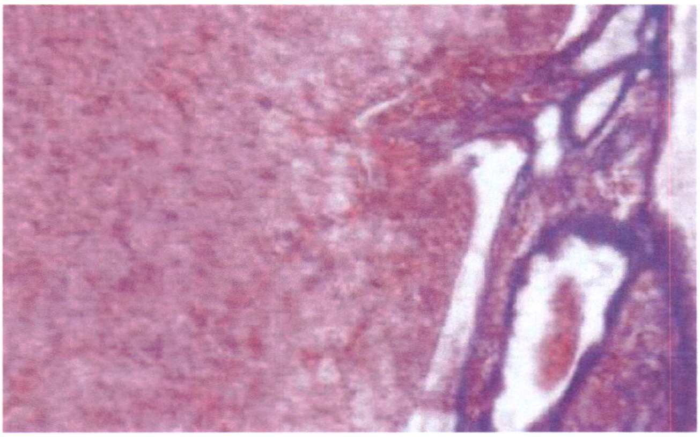

Fig. (8). T.S of *Channa punctatus* infected pancreas tissue shows the weakly acidic sulfated mucosubstances (Aldehyde fuchsine).

Proteins

Mercuric Bromophenol Blue

The control pancreas tissue revealed great positivity for Bromophenol blue, indicating a high concentration of total protein. The hepatocyte cytoplasm displayed a positive reaction characterized by fine granulation uniformly dispersed. The hepatocytes and pancreatic tissue infected with bacteria and fungi showed moderate positivity. Total protein demonstrated a significant reduction in both the cytoplasm and nucleus of the hepatocytes and pancreatic tissue.

The histochemical methods employed in this study have unequivocally demonstrated that both fish species possess a substantial quantity of proteinaceous material in their pancreas. This was corroborated when pancreatic sections were subjected to additional specific stains for the identification of various proteins (Plate **XXXI**, Figs. **1** to **4**).

PLATE – XXXI

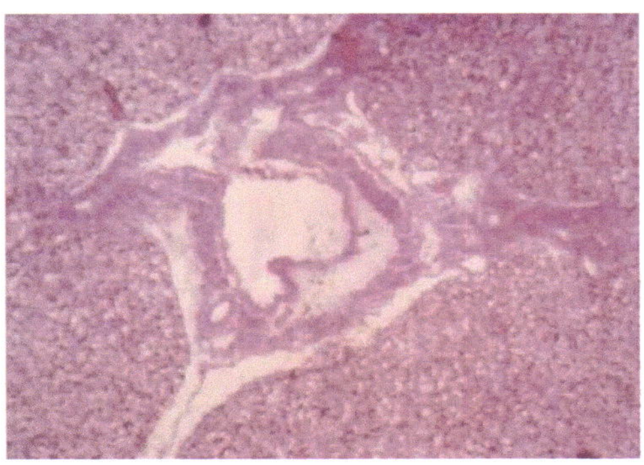

Fig. (1). T.S of *Channa striatus* control pancreas tissue displaying the total basic protein substances (Bromophenol blue).

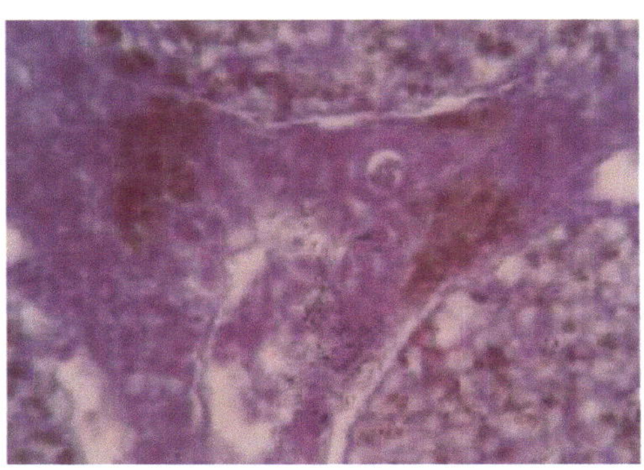

Fig. (2). T.S of *Channa striatus* infected pancreas tissue showing the basic proteins (Bromophenol blue).

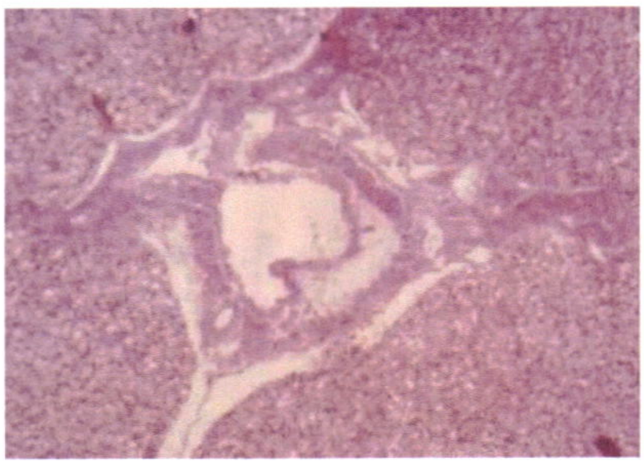

Fig. (3). T.S of *Channa punctatus* control pancreas tissue displaying the total basic protein substances (Bromophenol blue).

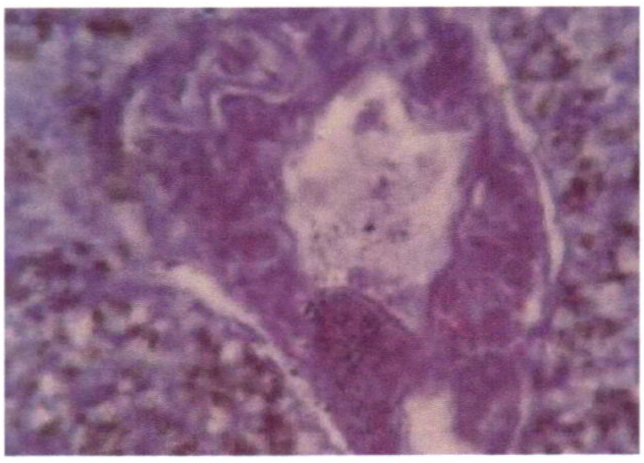

Fig. (4). T.S of *Channa punctatus* infected pancreas tissue showing the basic proteins (Bromophenol blue).

Congo Red

The pancreatic control cells exhibit significant positivity to Congo red, indicating the presence of protein-bound amino groups and glycoproteins within the cells.

The EUS-infected pancreas treated with Congo red exhibited a fairly positive reaction, indicating a total absence of protein, with remains primarily situated at the cell peripheries, which displayed cytoplasmic vacuolation (Plate **XXXII**, Figs. **1** to **4**).

PLATE – XXXII

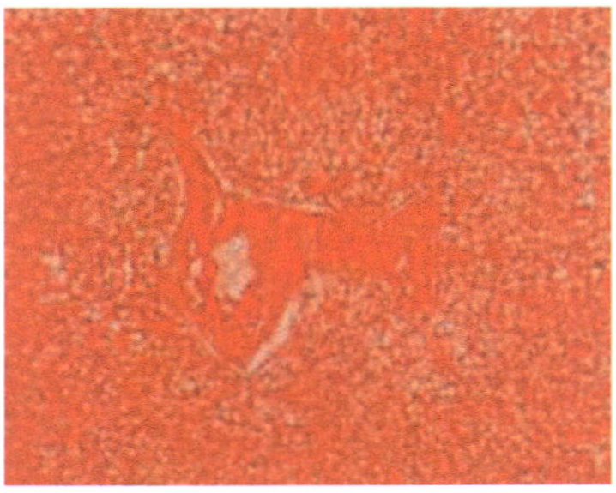

Fig. (1). T.S of *Channa striatus* control pancreas tissue showing protein-bound amino groups and glycoproteins (Congo red).

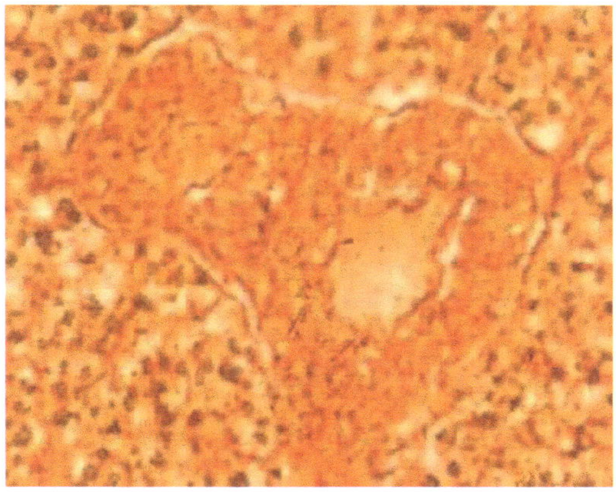

Fig. (2). T.S of *Channa striatus* infected pancreas tissue showing depletion of protein (Congo red).

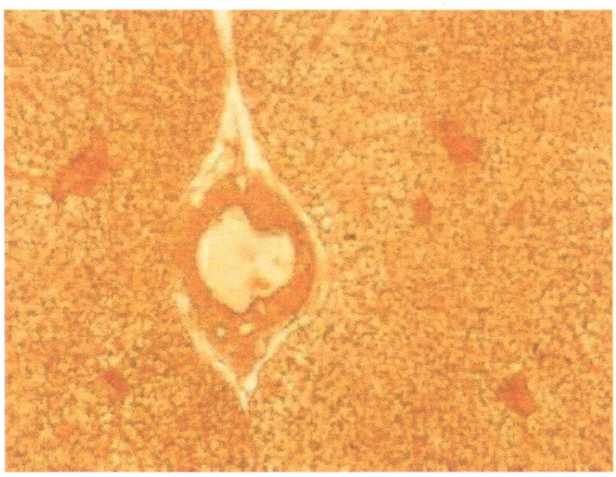

Fig. (3). T.S of *Channa punctatus* control pancreas tissue showing protein-bound amino groups and glycoproteins (Congo red).

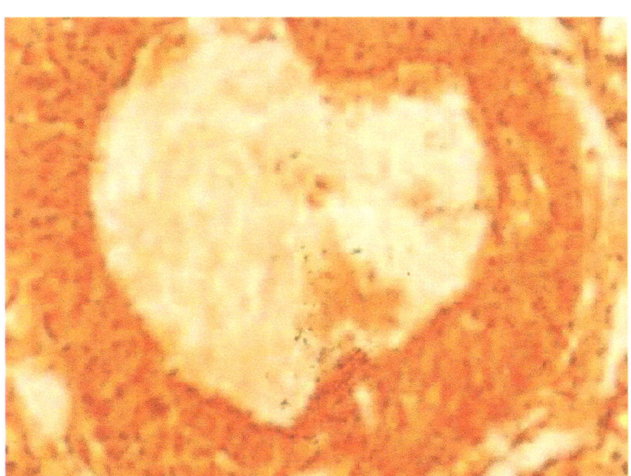

Fig. (4). T.S of *Channa punctatus* infected pancreas tissue showing depletion of protein (Congo red).

Fats

Sudan Black B

The pancreatic tissue has a robust reaction to Sudan black B, indicating the presence of lipids and phospholipids. Lipoid inclusions are uniformly dispersed within the cytoplasm, positioned perinuclearly and peripherally, with a notable accumulation in the cytoplasm close to the sinusoid.

The pancreatic tissues infected by bacteria and fungi were treated with Sudan Black B, resulting in a mild reaction that indicates a depletion of lipids and phospholipids. It disclosed an irregular and coarse lipid particle aggregation unevenly dispersed throughout the hepatic parenchyma. Conversely, the normal hepatopancreas had a uniform distribution of lipids inside the liver cells, evidenced by the presence of fine black cytoplasmic granulation (Plate **XXXIII**, Figs. **1** to **4**).

PLATE – XXXIII

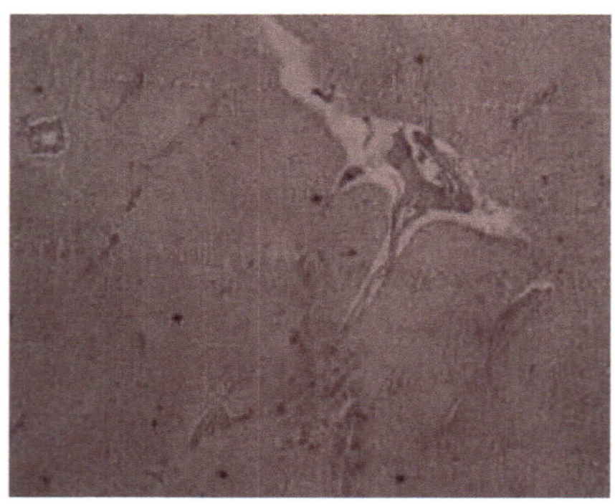

Fig. (1). T.S of *Channa striatus* control pancreas tissue exhibiting lipids and phospholipids (Sudan black B).

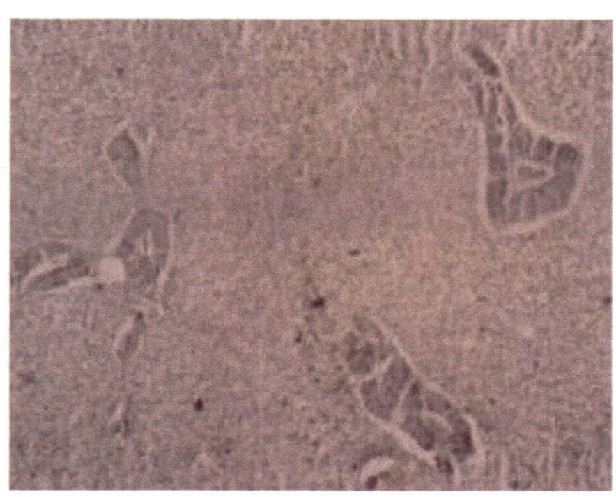

Fig. (2). T.S of *Channa striatus* infected pancreas tissue displaying depletion of lipids andphospholipids (Sudan black B).

Fig. (3). T.S of *Channa punctatus* control pancreas tissue exhibiting lipids and phospholipids (Sudan black B).

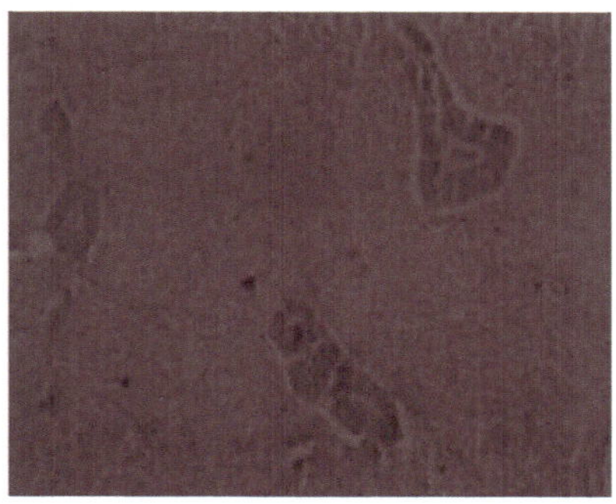

Fig. (4). T.S of *Channa punctatus* infected pancreas tissue displaying depletion of lipids andphospholipids (Sudan black B).

Ninhydrin/Schiff

The existence of protein-bound amino groups was confirmed using Ninhydrin/Schiff, yielding a moderately positive result in the control pancreatic tissue and a substantially positive result in the EUS-infected pancreatic tissue (Plate **XXXIV**, Figs. **1** to **4**).

PLATE – XXXIV

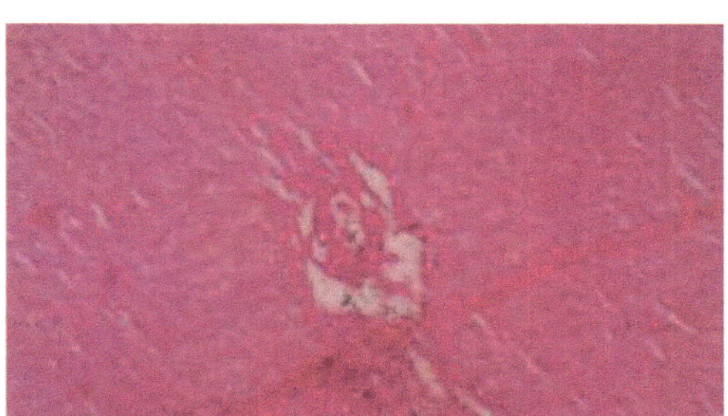

Fig. (1). T.S of *Channa striatus* control pancreas (Ninhydrin/Schiff).

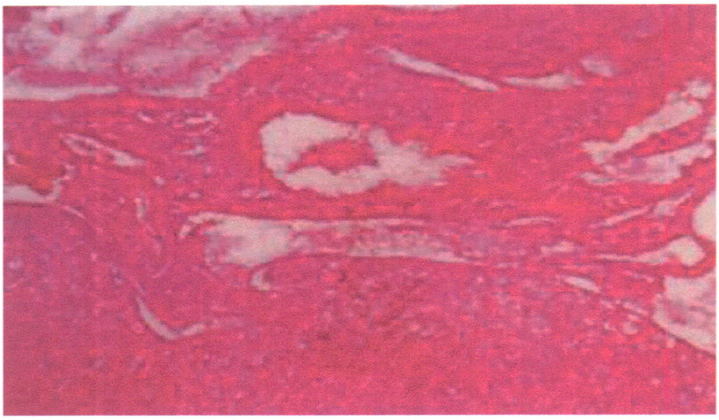

Fig. (2). T.S of *Channa striatus* infected pancreas (Ninhydrin/Schiff).

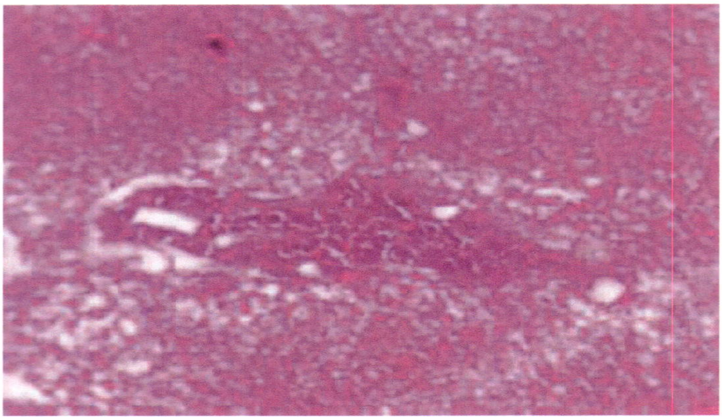

Fig. (3). T.S of *Channa punctatus* control pancreas (Ninhydrin/Schiff).

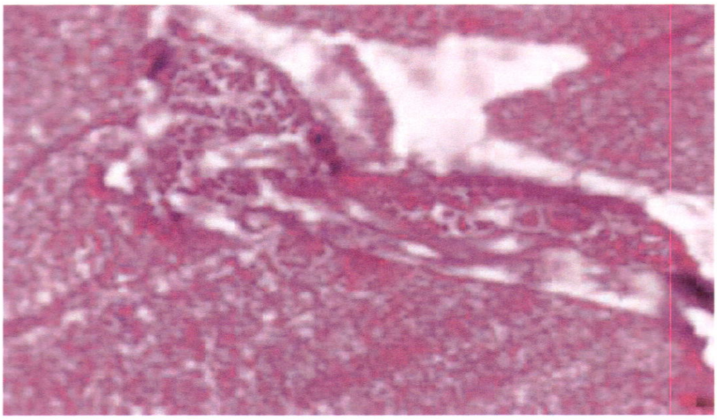

Fig. (4). T.S of *Channa punctatus* infected pancreas (Ninhydrin/Schiff).

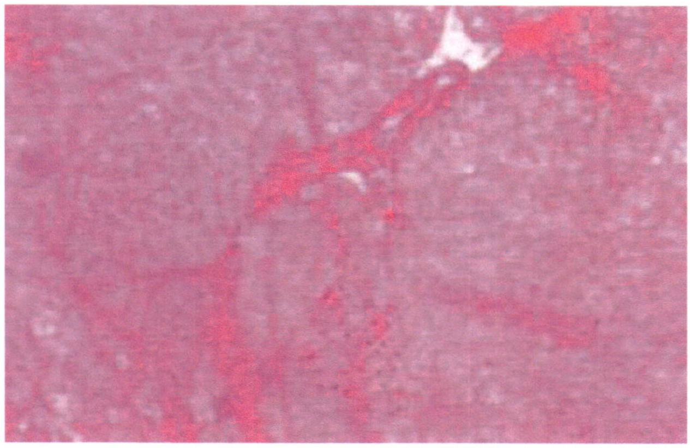

Fig. (5). T.S of *Channa striatus* control pancreas (Millon's reaction).

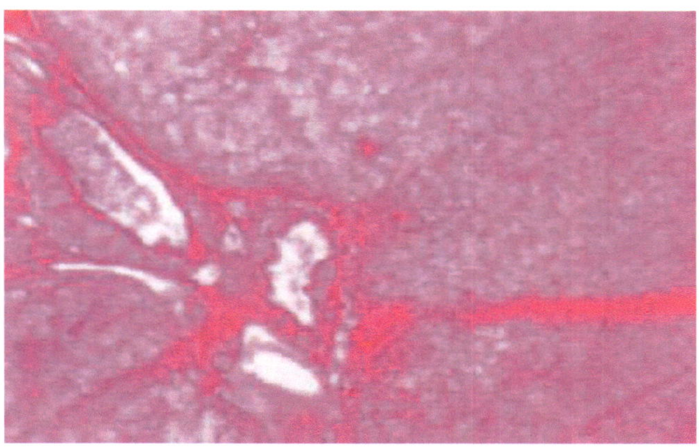

Fig. (6). T.S of *Channa striatus* infected pancreas (Millon's reaction).

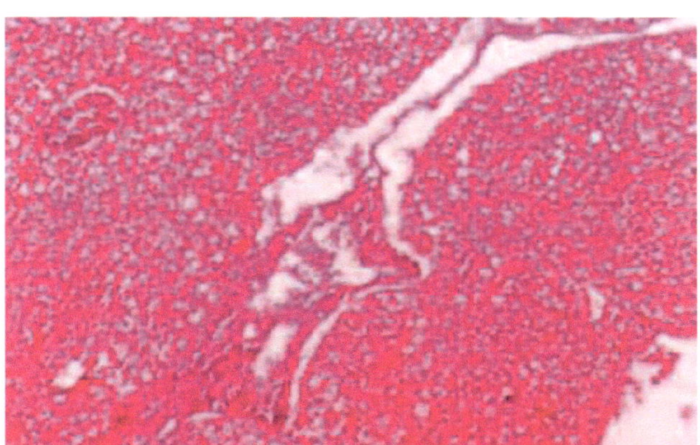

Fig. (7). T.S of *Channa punctatus* control pancreas (Millon's reaction).

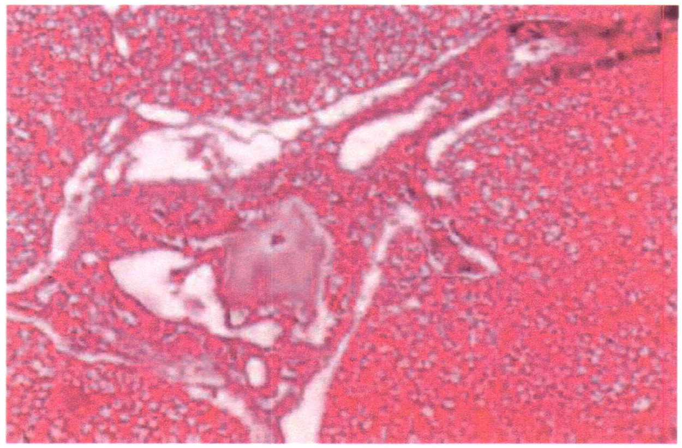

Fig. (8). T.S of *Channa punctatus* infected pancreas (Millon's reaction).

PLATE – XXXV

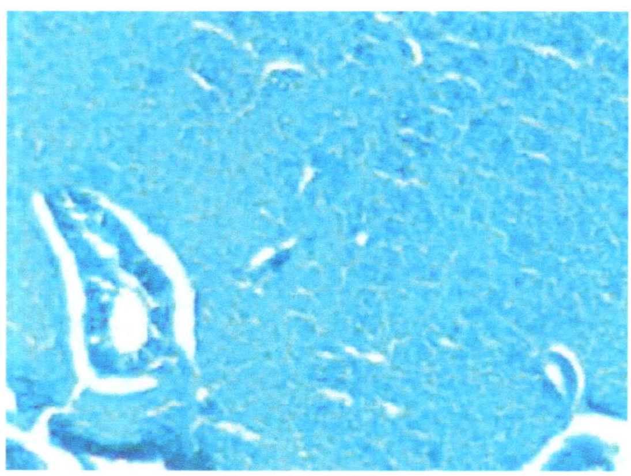

Fig. (1). T.S of *Channa striatus* control pancreas (FFC).

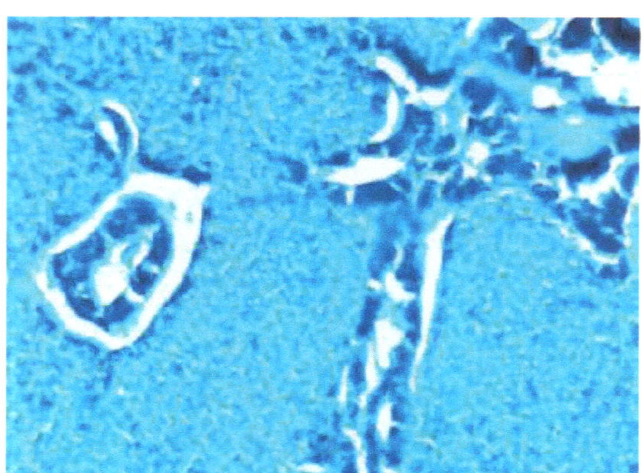

Fig. (2). T.S of *Channa striatus* infected pancreas (FFC).

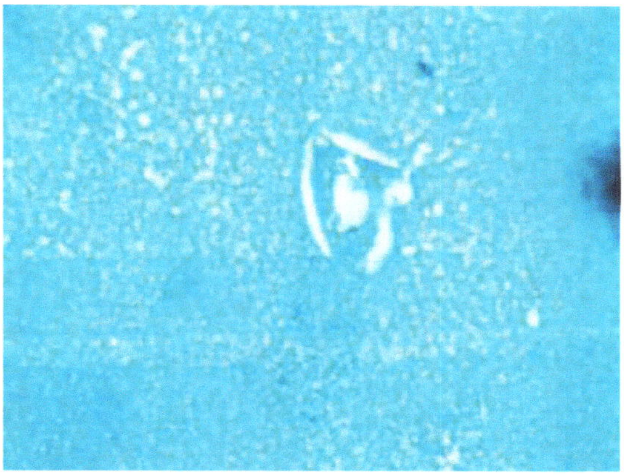

Fig. (3). T.S of *Channa punctatus* control pancreas (FFC).

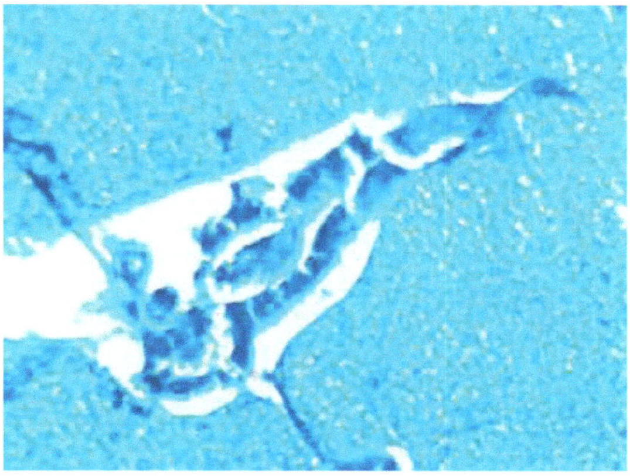

Fig. (4). T.S of *Channa punctatus* infected pancreas (FFC).

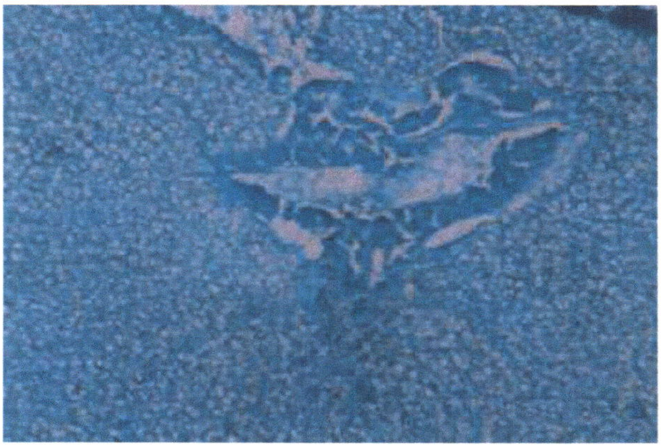

Fig. (5). T.S of *Channa striatus* control pancreas ($KMnO_4$/Alician Blue).

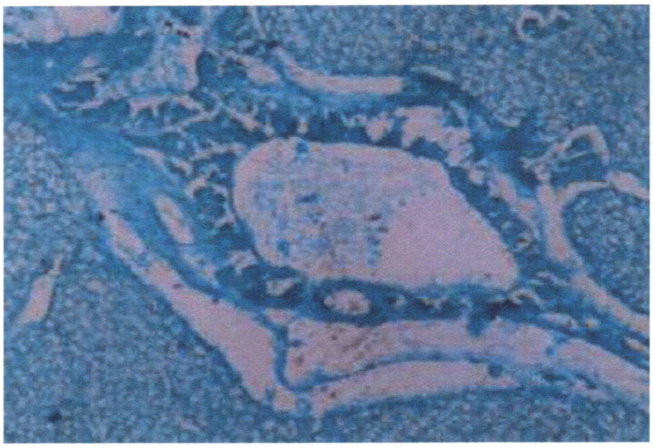

Fig. (6). T.S of *Channa striatus* infected pancreas ($KMnO_4$/Alician Blue).

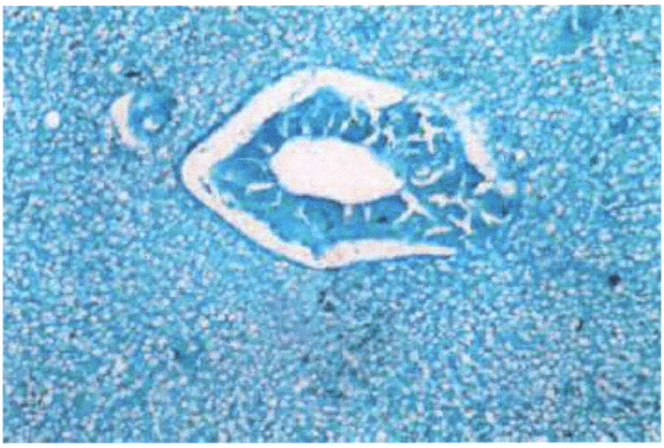

Fig. (7). T.S of *Channa punctatus* control pancreas ($KMnO_4$/Alician Blue).

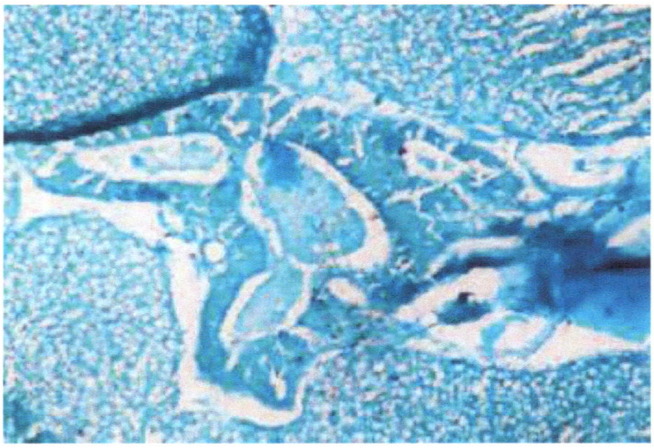

Fig. (8). T.S of *Channa punctatus* infected pancreas ($KMnO_4$/Alician Blue).

Millon's Response

Likewise, the amino acid tyrosine exhibited a somewhat positive response in the control pancreatic tissue and a positive response in the EUS-infected pancreatic tissue, as indicated by Millon's reaction (Plate **XXXIV**, Figs. **5** to **8**).

Ferric Ferricyanide (ffc)

The control pancreatic tissue exhibits modest levels of sulfhydryls, and disulfides were also identified. In the EUS, infected pancreatic tissue exhibited a favorable response to ferric ferricyanide (Plate **XXXV**, Figs. **1** to **4**).

$KMnO_4$/AB

The pancreatic tissue control exhibits modest positivity for cystine when stained with $Kmno_4$/Alcian blue. In the EUS, infected pancreatic tissue had a favorable response to $KMnO_4$/Alcian blue (Plate XXXV, Figs. **5** to 8).

Copper Phthalocyanine (CPP)

The pancreas, similar to other digestive glands such as the liver, contains carbohydrates and substantial quantities of proteins. The pancreatic tissue displayed a substantial presence of phospholipids and lipids. The histochemical assays performed on the hepatocytes of *C. Striatus* and *C. Punctatus* demonstrated the existence of substantial quantities of phospholipids exhibiting high positivity in control liver tissue when stained with Copper phthalocyanine. In the EUS, infected liver tissue has a modest positive reaction to Copper phthalocyanine. Consequently, the presence of several protein types is regarded in the pancreas as essential for the gland's particular dynamic protein activity (Plate **XXXVI**, Figs. **1** to **4**).

PLATE – XXXVI

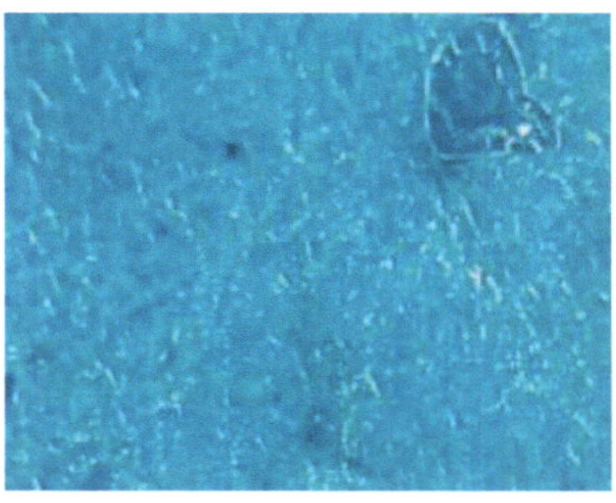

Fig. (1). T.S of *Channa striatus* control pancreas (CPP).

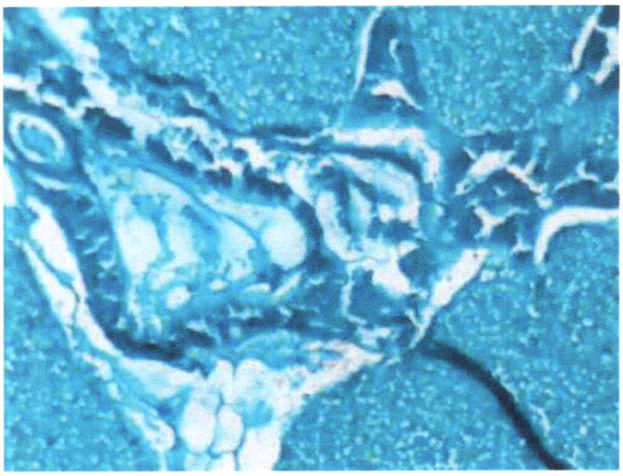

Fig. (2). T.S of *Channa striatus* infected pancreas (CPP).

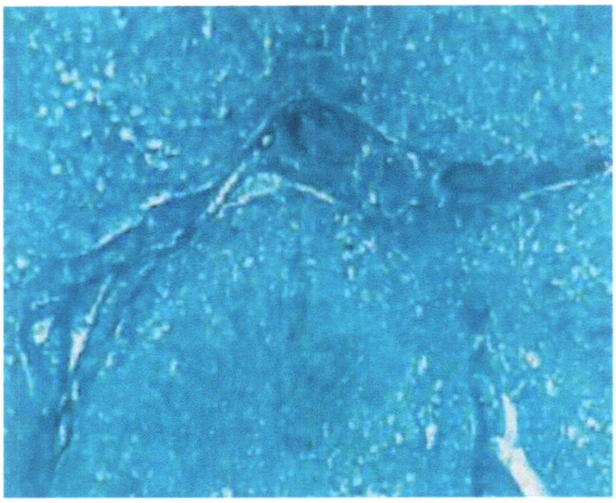

Fig. (3). T.S of *Channa punctatus* control pancreas (CPP).

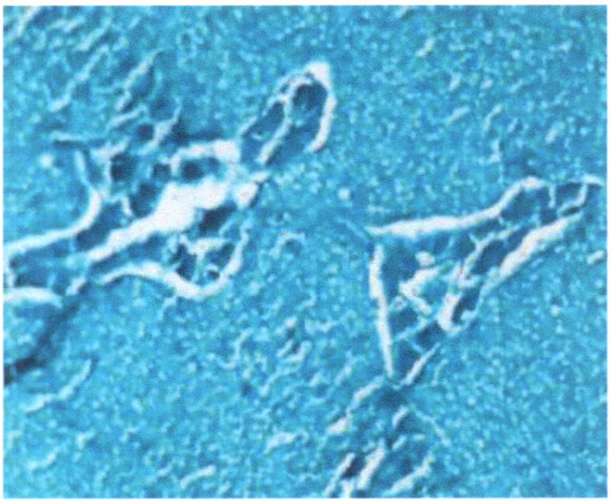

Fig. (4). T.S of *Channa punctatus* infected pancreas (CPP).

Nucleic Acids

Pyronin Y

The pancreatic tissue has a high reaction to Pyronin Y, indicating the presence of extracellular RNA in the control tissue sections.

The pancreatic tissues infected by bacteria and fungi were treated with Pyronin Y, resulting in a moderate reaction that indicates a depletion of RNA nucleic acids. This signifies a total absence of nucleic acids, with remnants primarily found at the cell peripheries, particularly in the hepatic parenchyma cells (Plate **XXXVII**, Figs. **1** to **4**).

PLATE – XXXVII

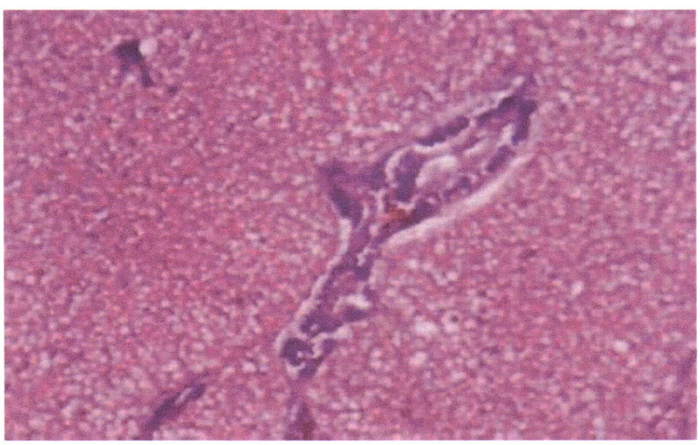

Fig. (1). T.S of *Channa striatus* control pancreas (Pyronin Y).

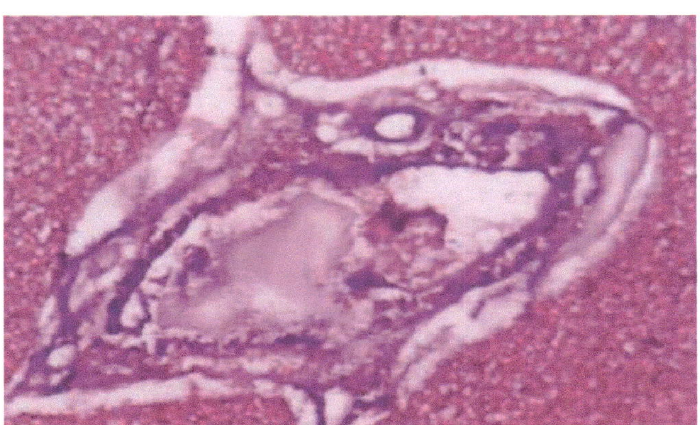

Fig. (2). T.S of *Channa striatus* infected pancreas (Pyronin Y).

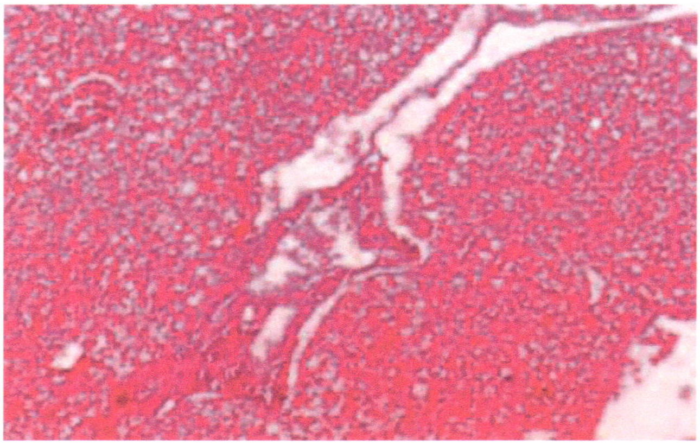

Fig. (3). T.S of *Channa punctatus* control pancreas (Pyronin Y).

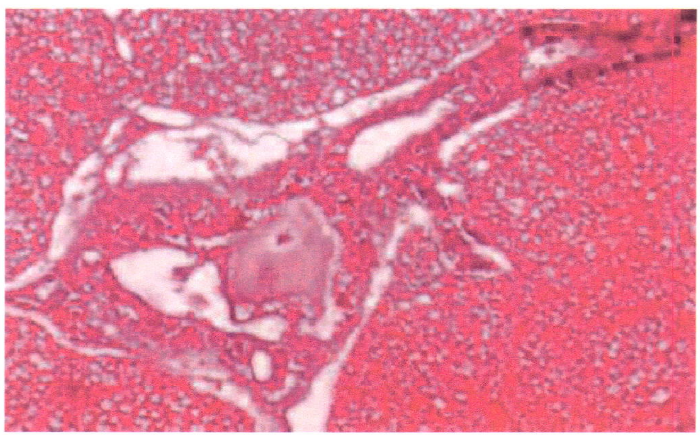

Fig. (4). T.S of *Channa punctatus* infected pancreas (Pyronin Y).

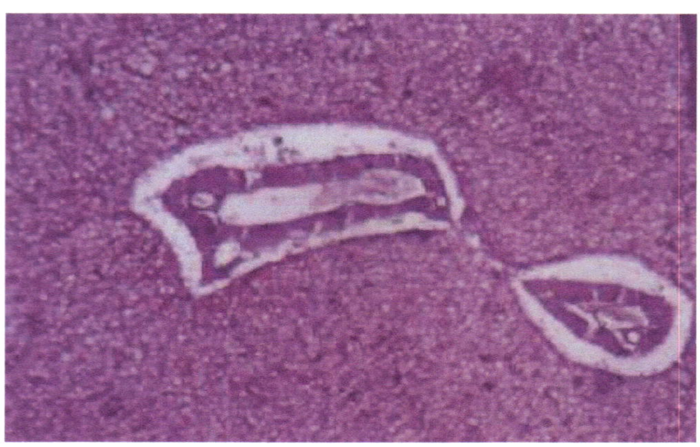

Fig. (5). T.S of *Channa striatus* control pancreas (Feulgen Reaction).

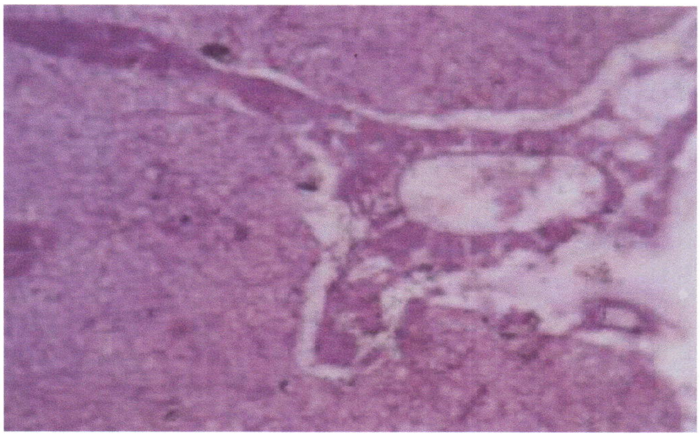

Fig. (6). T.S of *Channa striatus* infected pancreas (Feulgen Reaction).

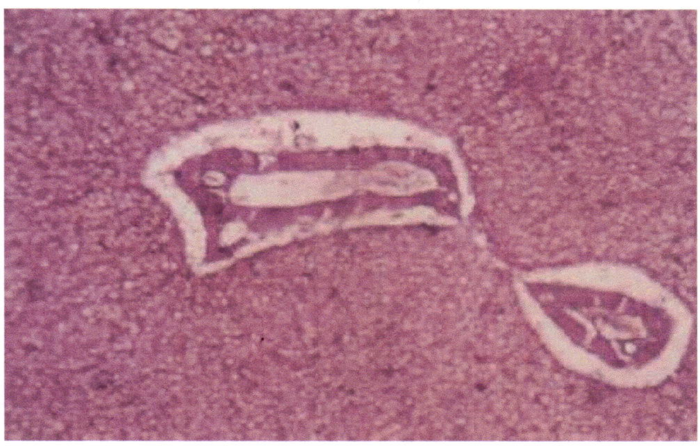

Fig. (7). T.S of *Channa punctatus* control pancreas (Feulgen Reaction).

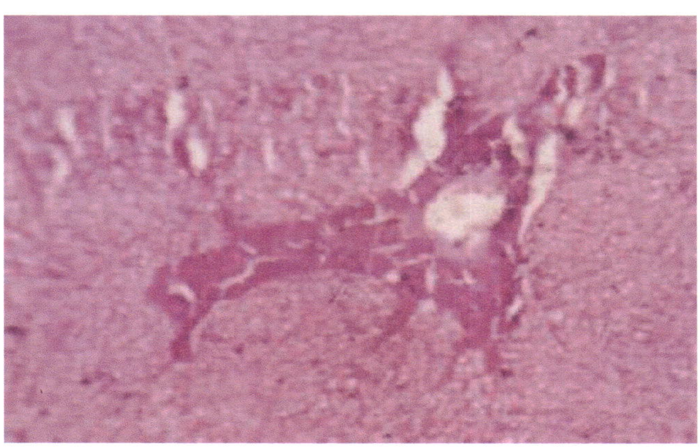

Fig. (8). T.S of *Channa punctatus* infected pancreas (Feulgen Reaction).

Feulgen Reaction

The pancreatic tissue exhibits a robust response to the Feulgen reaction, indicating the presence of extracellular DNA in the control tissue sections. The pancreatic tissues infected by bacteria and fungi underwent the Feulgen reaction, exhibiting a moderate response that indicates a depletion of DNA nucleic acids. This is evidenced by the irregular and coarse aggregates of lipid material unevenly distributed throughout the hepatic parenchyma cells (Plate **XXXVII**, Figs. **5** to **8**). The findings from the aforementioned histochemical tests are compiled in Table **62**.

Dialogue

This work has yielded the following histochemical observations; nonetheless, in the control *C. striatus* and *C. punctatus*, the epidermal cells of fish were abundant in glycoproteins, and the mucous cells also contained substantial amounts of glycoproteins, as indicated by Bromophenol blue. In addition to glycoprotein, the cytoplasm of these cells also included neutral polysaccharides and acid mucosubstances containing both sulfated and carboxylated groups. The primary constituents of these cells were sulfated mucins. These observations were correlated with those of numerous authors. The skin mucus of certain fish species comprises mucopolysaccharides containing sulphated and carboxyl radicals (Roberts, 1978).

In the mucous secretion of fish, interspecific variations were observed. In Pleuronectes platessa, goblet cells contain sulfated groups (Retcher and Grant, 1968), whereas in other species, they contain glycoproteins abundant in sialic acid and/or sialosulphomucins. (Askawa 1970; Harris *et al.* 1973; Sarasquete *et al.* 1989). The mucous cell of *I. Koreensis* was characterized by acidic sulfomucins, along with some sialomucins. The acidic characteristics of the mucous cells were documented in various fish species, including Monopterus, Mastacembelus, Amphipnous, and Misgurnus, which possess strongly acidic sulfated mucopolysaccharides. Additionally, Mittal and Munshi (1971); Mittal and Banerjee (1974); Mittal *et al.* (1980); Park and Kim (1999 and 2000) reported similar findings in Heteropneustes, which contain weakly acidic sulfated mucopolysaccharides. This mucus shows a significant capacity to retain substantial quantities of water (Letterer, 1959; Rogers, 1961), and the mucus released by the skin of air-breathing fish may also serve to maintain the skin's clarity for respiration. The mucous cells of Ancistrus exhibit a substantial appearance that appears to offset their limited quantity, leading to the high mucogenicity reported (Tarun *et al.*, 2010). The skin of the scaleless catfish

(*Clarias batrachus*) secretes substantial quantities of mucus rich in glycoproteins (Laxma Reddy and Benarjee, 2011).

Only the synbranchoid fish, M. Albus, possesses substantial unicellular mucous glands, which inhibit desiccation and reduce friction during its digging (Liem, 1967). The mucous cell of *L. Mediadiposalis* is characterized as acidic and sulfomucins in nature, according to Park *et al.* (2003). Acidic sulphated mucopolysaccharides in mucous cells have also been documented in other cutaneously respiring fish such as Monopterus, Mastacembelus, Amphipnous, and Misgurnus. Consequently, it can be inferred that mucus serves as an adaptive modification essential for the survival of fish residing in tranquil, shallow rocky stream habitats, frequently confronting the challenges of total desiccation, and may also function to maintain skin clarity for respiration (Rogers, 1961; Mittal *et al.*, 1980). The lubricative function of mucus diminishes bodily friction in aquatic environments, facilitating swimming and safeguarding the body from abrasion during burrowing and nest excavation. Mucous serves as a barrier against many infections, inhibiting their colonization in the epidermis (Lewis, 1970; Mittal *et al.*, 1994). Moreover, the sulfate groups induce acidity of glycoproteins, which effectively inhibits bacterial and fungal invasion.

The flattened cells identified in the uppermost layers of the epidermis of *Channa striatus* and *Channa punctatus* may be regarded as epidermal mucous cells that, upon the secretion of mucus, adopt a flattened morphology. Elevated levels of acid mucopolysaccharides, especially sulfated mucins, were also seen in these cells. Histological and histochemical criteria suggest that basal cells, referred to as stratum germinativum by numerous authors, progressively move outward to become flattened. This signifies the conclusion of their lifespan. A hypothesis supporting this is presented by Mittal *et al.* (1977), who found that alkaline phosphatase activity was highest in the penultimate layer of the polygonal epidermal cells. This was associated with the formation of mucopolysaccharides (Mittal and Banerjee 1974a; 1974b; 1975). Mucous granules were identified in the epidermal cells of Notopterus notopterus, Bagarius bagarius, and Channa striata. Kitzen *et al.* (1968) observed analogous findings regarding the epidermal cells of Protopterus annectens. Yamada (1968) and Whitear (1970) identified electron-dense vesicles, likely mucous granules, within the epidermal cells of many fish species. The existence of mucous-rich epidermal cells as an adaptation to an aquatic lifestyle is likely corroborated, as mucous has been documented in the epidermal cells of amphibians (Parakkal and Matoltsy, 1964) and turtles (Matoltsy and Huszar, 1972). However, such mucous granules have not been identified in the epidermal cells of terrestrial Squamata (Bryant *et al.*, 1967). The ovule cells of Chinese fish appear to be more histochemically related to flat cells than to those of mammals or birds. They appeared to produce mucous abundant in sialic acid, which included carboxylated mucous. However, the blue-purple hue exhibited by numerous cells upon staining with Aldehyde fuchsin/Alcian blue indicates that these cells also contained significant amounts of sulfated mucopolysaccharides, in contrast to the flat cells.

The mucous gland cells exhibited a positive reactivity to Bromophenol blue and Congo red, indicating the presence of sialic acid-containing glycoproteins. Flat cells had positive reactions solely to AB, indicating the presence of sialomucins, but not glycoproteins, as they did not react to either Bromophenol blue or Congo red. Histochemical analysis has confirmed the existence of sialic acid-containing glycoproteins in the dermal unicellular mucous gland cells of various fish species, including Salmo trutta (Harris, Watson, and Hunt, 1973), Malopterus electricus, and Torpedo ocellata (Carmignani and Zaccone, 1974 and 1975). Salvelinus alpinus (Pickering and Macey, 1977). Biochemically, glycoproteins containing sialic acid have been found on the surface mucous of several fish (Wessler and Werner, 1957; Fletcher and Grant, 1968; Asakawa, 1973). Another notable characteristic of these cells is their substantial glycogen content, as demonstrated by the saliva digestion/PAS technique. Glycogen, an important energy source, may play a role in the creation of mucopolysaccharides within these cells. Comparable findings have been documented in the developing trout Porcelli and Massari (1969), Salmo iridens Massari (1971), and Torpedo ocellata (Carmignani and Zaccone, 1975).

The secretion of epidermal mucous cells primarily comprises mucopolysaccharide-protein complexes with elongated structures. The facilitated movement of water over these long-chain molecules has been emphasized (Fabula *et al.*, 1966). Polypeptide-like organic K detritus on ship keels was discovered to diminish drag, hence preventing ships from achieving higher speeds (Barnaby and Dorecy, 1965). Uskowva *et al.* (1976) found a direct association between the protein content of the mucous and its friction-reducing potential. Consequently, the lowering of the frictional forces between the fish and water may be one of the most significant functions of the mucus produced by the epidermal mucous cells. They appeared to synthesize glycoproteins, as indicated by their responses with PAS, Bromophenol blue, and Congo red. The mucosal lining of the canal organ and the cupula contained acid mucopolysaccharides, predominantly sulfated mucins. The cytoplasm of the supporting cells exhibited analogous histochemical characteristics and may thus be somewhat accountable for the synthesis of the mucous lining and the cupula. No favorable response to AB has been observed in the canal wall cells surrounding the macula at either pH 2.5 or pH 1.0. Nevertheless, the cells of the remaining canal wall exhibited a marginal spositive to AB pH 2.5. Comparable observations have been noted by Bergeijk and

Alexander (1962) in the lateral line canals of Fundulus heteroditus. Mucous cells were detected throughout the canal, excluding the area surrounding the macula.

The glycoprotein composition of epidermal secretions in fish may differ according to species and cell types, as noted by Whitear (1986). In *Sparus aurata*, the progression of lymphocystis disease revealed the presence of proteins abundant in various amino acids inside the cytoplasm, nucleus, nucleoli, intracytoplasmic inclusions, and hyaline capsule (Gonzalez *et al*., 1996). Mucoproteins, glycoproteins, or acid mucopolysaccharides (sulfated and carboxylated) were identified in the hyaline capsule of lymphocystis-infected cells in various fish species (Nigrelli and Ruggieri 1965; Howse and Christmas 1978). The primary constituent of the mucus generated by goblet cells comprises weakly acidic mucins, which exhibit sialic acids and sulfated components with a reduced degree of sulfation, as demonstrated by suitable histochemical techniques.

In infected fish, glycogen was absent in these cells; however, basic proteins and trace amounts of proteins containing disulfides and sulfhydryls indicated the presence of glycoproteins in the surface slime of species such as Pleuronectes platessa. Fletcher and Grant (1968), Anguilla japonica Asakawa (1973), and Allolepis hollandi. Hashimoto and Yoshinura (1977) conducted histochemical demonstrations in the basal and epidermal mucous cells of Torpedo ocellata (Carmignani and Zaccone, 1975). These authors have indicated a direct association between protein concentration and the hydrodynamic efficacy of the mucus.

The typical fish mucous glands demonstrated significant positive results, indicating the presence of lipids. The findings of Mittal *et al*. (1976) indicate a correlation between elevated levels of cholesterol, its esters, and phospholipids with the metabolically active condition of basal cells engaged in proliferation and differentiation. Neutral lipids and phospholipids in the polygonal cells of the outermost layer may contribute to surface slime and serve as an effective barrier against water penetration through the skin.

The deposits of neutral lipids in the subcutis may supply energy during fasting, function as a barrier to water diffusion through the skin, and serve as shock-absorbing pads that protect the fish from mechanical harm. (Mittal and Banerjee, 1975) observed polygonal cells in the topmost layer, exhibiting pronounced succinic dehydrogenase activity and synthesizing sulfated acid mucopolysaccharides and lipids. A dense layer of slime composed of mucopolysaccharides, lipids, and fundamental proteins is likely crucial for maintaining skin moisture for cutaneous respiration, reducing water loss through evaporation, aiding in burrowing in mud and swimming in water, and safeguarding the skin against bacterial and fungal infections.

The current investigation revealed an increase in carbohydrate levels in the infected fish. These observations aligned with the findings of Zaccone *et al.* (1985), who detailed the histochemistry of several oxidative enzymes and complex polysaccharides in the epidermis of the catfish Heteropneustes fossils following exposure to sublethal amounts of the detergent sodium alkylbenzenesulphonate.

Consequently, our observations have revealed substantial evidence of specialization in the integument of Channa striatus and *Channa punctatus*. Aside from these studies, there is minimal information regarding the histochemical characteristics of infected fish. Nonetheless, greater investigation into the etiology and other facets of fish infestations remains necessary.

The control teleost fishes have two sets of four holobranchs that constitute the lateral aspects of the throat. Each holo-branch comprises two hemibranchial structures extending from the posterior margin of the branchial arch. The respiratory lamellae possessed a unique epithelial layer and blood capillaries. The interlamellar areas were distinct and pronounced.

The histochemical analyses conducted in the current microscopic examinations of the gills demonstrated alcianophilia of the mucus glands when stained with alcian blue at pH 1.0 and pH 2.5. The staining intensity at pH 1.0 was less pronounced compared to that at pH 2.5. This distinctly indicated the presence of sulphated mucosubstances, hyaluronic acid, and sialic acid in the gills. Alcian blue (1.0 pH)/PAS revealed the presence of acid mucopolysaccharides, sulfated mucosubstances, hyaluronic acid, and sialomucins. The application of alcian blue (2.5 pH)/PAS stain demonstrated the existence of neutral and strongly acidic mucopolysaccharides. The arsenic salt exhibited a significant toxic effect on the gills of C. Striatus and C. Punctatus, which exhibit periodic fluctuations in density, area occupancy percentage, and staining characteristics of the mucous cells. The substantial secretion of sulphated mucosubstances by the mucous cells has been reported in the literature, revealing that during severe stages of infection, there was a considerable reduction in both the density and area occupancy % of the gill mucous cells under seasonal conditions. Nevertheless, the density persisted above the usual level until the end of the seasonal period, while exhibiting periodic changes. This suggests the regeneration of a substantial number of small-sized mucous cells. Subsequently, these indicators exhibit fluctuations at various periods of exposure and consistently remain subnormal, particularly during the later stages of EUS infection. Bacteria and fungi have irrevocably modified the monogenic function of gill mucous cells, as seen by the reduced density and area occupancy of these cells. Excessive mucus coagulation on respiratory surfaces may disrupt several critical physiological processes, including gas exchange, nitrogen excretion, salt balance, and blood circulation (Laurent and Dunel, 1978). Additionally, it temporarily inhibits the penetration of ambient arsenic salts.

The gill architecture in teleost fishes is likely to be significantly altered, leading to pathological conditions when exposed to environmental and chemical pollutants such as heavy metals, pesticides, and insecticides (Benarjee, 1986). These alterations are particularly evident at the base of the secondary gill lamellae and the respiratory epithelium of both primary and secondary gill lamellae, with necrosis of the lamellar epithelium resulting in respiratory and osmoregulatory distress (Skidmore and Tovell 1972; Smith and Piper 1975; Lamke and Mount 1963; Christae and Battle 1963; Ashley, 1970).

The current observations indicate that bacterial and fungal agents have similarly infected *C. Striatus* and *C. punctatus*; however, the extent of harm inflicted was exacerbated by the prolonged duration of seasonal illness. The primary alterations included necrosis, capillary rupture, and atrophy of respiratory gill filaments. The outcomes of the current experiments demonstrate a resemblance to the discoveries made by Rao *et al.* (1983).

Extensive secretion of mucus in fish acutely exposed to arsenic has been reported to result in mortality, as increased mucus production leads to suffocation or direct adverse effects on the gill epithelium. A review of arsenic toxicity indicates that arsenic-exposed fish experience respiratory difficulties due to gill obstruction by coagulated mucus, vascular collapse in the gills, and anoxia resulting from the direct damaging effects of arsenic ions on blood vessels (Irwin, 1997). Prolonged exposure of *Clarias batrachus* to sub-lethal concentrations of sodium arsenate did not result in mortality, despite considerable mucous discharge observed in all respiratory organs, including the epidermis (Singh and Banerjee, 2008a). Comparable mucus discharges from the respiratory organs of several fish species have often been observed following exposure to multiple heavy metal salts (Rajan and Banerjee, 1991; 1992; Hemalatha and Banerjee, 1993; Parashar and Banerjee, 1999a, b, c; Banerjee and Chandra, 2005). The authors did not observe any fish mortality resulting from mucus coagulation. Mortality of fish observed (Sorenson *et al.*, 1979). Additionally, some effects may result from damage inflicted by arsenic salts on numerous cellular components of important organs, including the gills of fish (Singh *et al.*, 2014).

The liver of freshwater teleosts is a crucial digestive organ composed of two primary lobes. Each lobe consists of polyhedral cells organized radially around the hepatic vein in ductal formations. Sinusoids and intracellular bile canaliculi are dispersed among the hepatocytes.

The microscopic examination of the liver parenchyma indicated that the histochemical changes were not confined solely to the infection sites but also affected various regions of the organ due to parasite replication, invasion, and the secretion of certain toxins, which may lead to disruption, tissue damage, and a subsequent reduction in immune response, thereby exacerbating the histochemical alterations in the infected animals.

This study performed a range of histochemical tests to elucidate the chemical makeup of liver cells. Histochemical assays indicate that liver cells contain substantial quantities of carbohydrates, proteins, lipids, and nucleic acids. The PAS positivity indicates the presence of carbohydrates, but the alcian blue test yielded a negative result, signifying the absence of mucopolysaccharides. Proteins were abundantly characterized by substantial amounts of glycoproteins and moderate levels of tyrosine, tryptophan, sulfhydryls, and disulfides. Phospholipids and lipids were additionally detected. The Pyronin Y and Feulgen reactions indicated the presence of RNA and DNA.

Environmental pollutants, low oxygen levels, low temperatures, and bacterial and fungal pathogens prevalent in freshwater infiltrate fish through ingested water, the oral mucosa, or gills, potentially reaching the liver *via* blood circulation. Structural differentiation and functional alterations in the liver are induced by the bacteria *Aeromonas hydrophila*, *Staphylococcus aureus, Pseudomonas aeruginosa, Salmonella salmonicida*, and the fungi *Aspergillus fumigatus, Trichoderma viride, Aspergillus niger, Rhizopus stolonifer, Penicillium chrysogenum, Fusarium solani,* and *Trichoderma harzianum*.

Significant quantities of glycogen and lipids were identified in the hepatic tissue using histochemical analysis. This aligns with the findings of Scarpelli *et al.* (1963) and Weis (1972), who noted a glycol and protein ratio in *Channa striatus* and *Channa punctatus*, indicating that the liver and pancreas primarily function to store glycogen.

The current study indicates that vacuoles within the cytoplasm of hepatocytes may harbor lipids and glycogen, which are associated with the liver's normal metabolic function. Similar findings were reported by Seyrafi *et al.* (2009 and 2011), who noted the emergence of vacuolar structures in hepatic cells, likely attributable to lipid presence, and identified substantial glycogen deposits at the cellular level through PAS reactions. Glycogen depletion in hepatocytes is typically observed in diseased animals. Hinton and Lauren (1990) and Wilhelm Filho *et al.* (2001) indicate that glycogen serves as a glucose reserve to meet the increased energy demands in such circumstances (Panepucci *et al.*, 2001).

The current study demonstrated a considerable reduction in hepatic total protein levels in infected fish, as indicated by the BPB technique. These findings align with those of Tripathi and Verma (2004), who indicated that exposure of *Clarias batrachus* to fenvalerate resulted in a substantial reduction of protein levels in the liver, brain, and skeletal muscle. Reddy *et al.* (1991) observed a decline in total, structural, and soluble proteins, while free amino acids and the activities of protease, aspartate aminotransferase, and alanine aminotransferase significantly increased in fenvalerate-exposed *Cyprinus carpio.*

The current study demonstrated a considerable reduction in carbohydrate content in infected fish, as indicated by the PAS technique. Singh and Srivastava (1981) indicated that carbohydrates diminished due to exposure to sublethal concentrations of a mixture of aldrin and formothion. Krapagaganapthy *et al.* (1988) reported a progressive decrease in the carbohydrate content of the liver in Colisa lalia, as evidenced by reduced staining intensity with PAS. They also proposed that carbohydrates in the liver are utilized more extensively under stressful conditions.

Sakr and Jamal Allail (2005) revealed a significant drop in liver cell glycogen levels compared to normal fish, noting that these alterations were time-dependent. Naik *et al.* (2004) showed a substantial reduction of total protein, glycogen, and lipid content in the tissues of *Cyprinus carpio* exposed to tannery effluent. Anandhi and Murthy (2006) investigated the localization of lipid, glycogen, and protein in the hepatic cells of *Glossogobius giuris* at various stages of its reproductive cycle. They discovered a significant reduction in protein, glycogen, and lipid levels in the livers of fish treated with Malathion.

The results indicate that the bacteria and fungi *Aeromonas hydrophila, Staphylococcus aureus, Pseudomonas aeruginosa, Salmonella salmonicida, Aspergillus fumigatus, Trichoderma viride, Aspergillus niger, Rhizopus stolonifer, Penicillium chrysogenum, Fusarium solani*, and *Trichoderma harzianum* induce a reduction in protein, lipid, and glycogen levels in the liver. The infected fish exhibit a significant decrease in the protein, lipid, and glycogen content in their tissues (Rogar, 1980).

A similar reduction in the levels of protein, lipid, and glycogen was reported by McLeay and Brown (1974); Sakr and Jamal Al Lail (2005); D'Cruz and Miranda (2006); Yadav *et al.* (2007); and Pathan *et al.* (2009). The current findings on *Channa striatus* and *Channa punctatus* suggest that the diminished levels of glycogen, protein, and lipid in the tissues may be attributed to contamination of the water by Aeromonas species and Aspergillus species or to stress imposed by microbial conditions.

The current study indicates that bacterial and fungal infections have resulted in fast degeneration, hypertrophy, necrosis, and vacuolation of hepatocytes, accompanied by tissue disintegration. Significant impacts were noted throughout the seasonal illness. This encompasses the loss of the distinctive hexagonal morphology of hepatic cells, rupture of nuclear and cellular membranes, creation of binucleate hepatocytes, increased atrophy at the center, and disarray of hepatic cords. The degree of liver damage is contingent upon seasonal fluctuations that contribute to EUS conditions.

The histochemical analyses in the current study indicate that pancreatic cells are abundant in carbohydrates and include modest levels of proteins such as tyrosine, tryptophan, sulfhydryls, disulfides, proteins with bound NH2 groups, and glycoproteins. They are also rich in lipid content. Shyamasundari *et al.* (2007) reported that the cytoplasm of pancreatic cells exhibits stronger staining with protein stains such as bromophenol blue or aldehyde fuchsine. The perimeter of the islets exhibited reduced intensity in staining with all histochemical stains, according to their research. Consequently, it is evident that cytochemical investigations in the pancreas are significantly constrained, necessitating further histological and cytological research in this area.

The exocrine pancreatic tissue comprises acinar cells akin to those found in teleostean species. Histochemically, glycogen, glycoproteins, protein-bound amino groups, and lipids must be present. Both exocrine and endocrine tissues exhibit a high concentration of RNA in their cells.

The majority of studies focus on understanding the function of the endocrine pancreas in freshwater fish, typically through the exogenous delivery of islets of Langerhans. The microscopic examination of the pancreas in freshwater fish is crucial for numerous essential activities. The current study demonstrates that the exocrine component of the pancreas in *C. Striatus* and *C. Punctatus,* and murrel, exhibits two forms: diffused and intrahepatic. The structure comprised densely arranged pancreatic acini displaying characteristics of protein-secreting cells, formed by pyramidal acinar cells with numerous zymogen granules and extensive rough Endoplasmic Reticulum (rER). Comparable findings were noted in Schilbe mystus and Labeo niloticus (Sheibani and Yali 2006; Huso huso and Naguib *et al.*, 2009).

In higher vertebrates, the pancreas is a compact and lobulated gland, while in lesser vertebrates, it exhibits significant variability. The anatomical and histological adaptations of the pancreas, whether in a diffuse state or as a compact gland, in relation to dietary habits, have been examined by various researchers. Khanna (1963) reported the infiltration of pancreatic acini into hepatic tissue, resulting in the formation of a compact gland referred to as the hepatopancreas. Gonzalez *et al.* (1993) documented the existence of a diffuse exocrine pancreas that extends through the mesentery, surrounding certain digestive organs or dispersing within the intraperitoneal adipose tissue in Serranus cabrilla. El-Said *et al.* (2010) similarly characterized the acinar configuration of the hepato-pancreas in *Mugil cephalus* and *Sparus aurata*. Nonetheless, the pancreas is structured as a compact gland similar to that of higher vertebrates, as documented by Kapoor (1953) and Khanna (1963) in the carnivorous species Wallago attu, Mystus aor, and Heteropneustes fossilis. In certain primitive fish such as Amia, Fundulus, and Oligosarcus jenynsii, the exocrine pancreas is situated as a diffuse organ that extends into the liver, forming a hepatopancreas (Boldyreff 1935; Petcoff *et al.*, 2006).

Seshadri (1961) documented a limited quantity of pancreatic acini in the carnivorous *Ophiocephalus striatus*, but he saw densely packed acini in the herbivorous Mugil speigleri. He observed a variation in the number of zymogen granules in these fish and attempted to correlate these variations with the differing feeding patterns of the analyzed species. The placement of the exocrine pancreas into the liver has also been documented by Petcoff *et al.* (2006) in the carnivorous teleost Oligosarcus jenynsii, Mrigala, and the intestine of niloticus is extensively coiled, an adaptation that facilitates prolonged retention of dietary material, as the sole source of pancreatic enzymes is the hepatopancreas. The pancreatic and bile juices, after secretion from the hepatopancreas, are stored in the gallbladder and then discharged into the alimentary canal (Chakrabarti *et al.*, 2012).

The zymogen granules house proenzymes that facilitate the digestion of proteins, carbohydrates, lipids, and nucleotides (Alboghohobeish and Khaksar-Mahabady, 2005). The acinar cells of the *C. Striatus* and *C. Punctatus* exhibited variations in the staining activity of zymogen granules using Bromophenol blue, signifying the presence of distinct enzymes within the granules. Enzyme histochemistry of the acini corroborates this, demonstrating varying ratios of distinct digestive enzymes, with elevated lipase and alkaline phosphatase activity and moderate acid phosphatase activity. The histochemical study of the pancreatic acini of the murrels revealed the presence of neutral mucopolysaccharides, glycogen, and lipid, indicating a high storage activity of these cells.

They are referred to by various names, including adipocytes, lipocytes, pericytes, and parasinusoidal cells (Wake, 1980). Ultrastructurally, they are identified by the presence of numerous lipid droplets within their cytoplasm and are present across a diverse array of species, from lampreys (primitive fish) to humans, having been observed in various tissues, including the liver, adrenal gland, spleen, ductus efferent, and uterus (Wake *et al.*, 1987; Geerts, 2001). These cells exhibited the unique filamentous protein marker α-smooth muscle actin, as noted by Yokoi *et al.* (1984) and Zhao and Burt (2007). Kordes *et al.* (2009) indicated that Pancreatic Stellate Cells (PSCs) are responsible for storing vitamin A and, upon injury, undergo transformation into an active state that generates substantial quantities of fibronectin and laminin (extracellular matrix proteins), leading to fibrosis. Shimizu *et al.* (2005) noted that these cells had an immunological activity since they engulfed injured pancreatic parenchymal cells. Rodlet cells were identified in the pancreatic acinar cells and the epithelium of the pancreatic duct of grass carp, as noted by Bielek (2005).

The distributions of the endocrine component of the pancreas, specifically the Langerhans islets, are situated between the exocrine acinar cells and surrounding the pancreatic duct (AZAN). The islet is encased by a delicate connective tissue capsule (AZAN). Oval-shaped alpha cells with big polyhedral beta cells (AZAN). Semi-thin sections of the islets of Langerhans exhibited oval-shaped alpha cells and massive polyhedral beta cells, as indicated by Bromophenol Blue staining. These may also be regarded as non-specific immune cells, participating in immunity due to their higher prevalence during parasite infections. Reite (2005) and Reite and Evensen (2006) classified them as a subtype of eosinophilic granulocyte. These cells are found in various organs of fish, including the heart, kidneys, gills, and intestines. Mokhtar and Abd-Elhafeez (2014) validated the secretory function of these cells ultrastructurally through a holocrine method of secretion. Melanomacrophage centers were identified in the exocrine region of the pancreas. The quantity and composition of melanomacrophage centers vary with species, health condition, and age (Agius, 1980).

They typically accumulate lipofuscin, melanin, ceroid, or hemosiderin pigments (Munshi and Dutta, 1996). Both centroacinar and duct cells evidently release bicarbonate and water, as carbonic anhydrase, the enzyme responsible for bicarbonate production, has been identified in their epithelia (Sheibani and Yali, 2006). The existence of goblet-like cells in the interlobular duct wall indicates an adaptation for pathogen defense, as these cells were also identified in common carp (Neuhaus *et al.*, 2007).

The robust PAS-positive apical segment of the major pancreatic duct suggested the presence of neutral mucopolysaccharides and may imply the transfer of molecules across the cell membrane. The primary duct of the grass carp pancreas discharged into the intestinal bulb, and its wall comprised argyrophilic cells that might potentially be somatostatin or glucagon-producing cells, as evidenced by their presence in the related species *Cyprinus carpio* (Kong *et al.*, 2002). The endocrine component of the murrel's pancreas was dispersed among the exocrine pancreatic tissue, the liver, and the intestinal bulb and intestine. These descriptions are analogous to those provided for islet aggregates in Salmoniformes (Wang *et al.*, 1986).

CONCLUSION

The Epizootic Ulcerative Syndrome (EUS) presents a considerable threat to global aquaculture, impacting numerous fish species. The effective management of EUS depends on a synthesis of precise diagnostic methods, preventive actions, and control strategies. Advanced diagnostic instruments, such as molecular methods like PCR, along with histopathological and histochemical analyses, have improved the early detection and surveillance of the disease. Preventive approaches, including enhancing water quality, regulating environmental stressors, and implementing biosecurity standards, are crucial in reducing epidemics. Molecular diagnostics and environmental monitoring are crucial for prompt response. Preventive interventions, such as water quality control, biosecurity measures, and stress alleviation, are essential in diminishing vulnerability to the disease. Control strategies, encompassing antifungal therapies, vaccine development, and the amalgamation of conventional and contemporary management techniques, are crucial for alleviating the effects of EUS. Ongoing research and collaboration among stakeholders, including fish farmers, researchers, and policymakers, are essential for promoting sustainable aquaculture methods. Through the implementation of comprehensive and proactive methods, the aquaculture sector can effectively address EUS, safeguarding fish health, minimizing economic losses, and fostering environmental sustainability. Continued investigation into disease processes, host-pathogen interactions, and innovative therapeutic strategies will facilitate enduring solutions to EUS. Epizootic Ulcerative Syndrome (EUS) remains a considerable menace to aquaculture, resulting in major economic detriment and impacting aquatic biodiversity. Effective management of EUS necessitates a comprehensive approach that includes advanced diagnostic methods, strong prevention strategies, and dependable control measures.

AI and ML-driven preventive techniques represent a significant transformation in aquaculture disease management. Predictive modeling facilitates the early detection of outbreak hazards, whereas optimization methods guarantee sustainable and efficient agricultural operations. Decision support systems facilitate equitable access to sophisticated technology, enabling farmers and aquaculture managers to proactively address EUS hazards. These innovations collectively improve resilience, productivity, and sustainability in global aquaculture systems. The management of Epizootic Ulcerative Syndrome (EUS) in aquaculture is on the verge of transformation through the implementation of Artificial Intelligence (AI) and Machine Learning (ML). These technologies provide robust solutions for the issues associated with disease detection, prevention, and control, hence maintaining the sustainability of aquaculture activities.

Revolutionizing EUS Management with AI and ML

Artificial Intelligence and Machine Learning facilitate a transformative change in Endoscopic Ultrasound management by implementing:

Improved Diagnostic Precision

AI-driven picture recognition and predictive models provide swift and dependable disease identification, exceeding conventional techniques in both speed and accuracy. Data-Driven Prevention, predictive analytics, and decision-support systems provide proactive interventions utilizing environmental and historical data, hence mitigating outbreak risks. Efficient Control Strategies: AI-driven treatment recommendation systems, together with continuous monitoring, enhance interventions while reducing resource consumption and environmental impact. These inventions enhance fish health and farm productivity while also addressing global issues such as food security and sustainable aquaculture. The significance of technology adoption. To harness the potential of AI and ML, it is essential to surmount obstacles to adoption and cultivate a collaborative ecosystem. Empowering small-scale farmers through accessible, cost-effective technologies tailored to their needs can democratize the advantages of AI, facilitating widespread adoption. Advancing Sustainability: AI technologies enhance resource efficiency, diminish waste, and lessen dependence on detrimental methods, harmonizing aquaculture with environmental conservation objectives, enhancing Resilience by incorporating real-time monitoring, early warning systems, and sophisticated analytics. Aquaculture operations may more effectively respond to problems such as climate change and disease outbreaks. The integration of AI and ML is essential for the future of sustainable aquaculture. By embracing these technologies, governments, researchers, technology developers, and farmers can advance the aquaculture industry towards a future where EUS and other illnesses are effectively managed, safeguarding the health of aquatic ecosystems and the lives of millions reliant on them. Artificial Intelligence and Machine Learning provide the means to harmonize productivity with sustainability, facilitating a robust and thriving aquaculture industry.

CHAPTER 5

Diagnostic Techniques Based on AI and Machine Learning

INTRODUCTION

Epizootic Ulcerative Syndrome (EUS) is a catastrophic disease impacting aquaculture, primarily caused by the fungal pathogen *Aphanomyces inva*dans, in conjunction with bacterial and viral diseases. It presents a considerable risk to global aquaculture, affecting the livelihoods of millions and the accessibility of sustainable seafood resources. The disease presents as ulcerative lesions, resulting in elevated mortality rates, diminished market value, and economic detriment for farmers. Conventional methods for controlling EUS depend on clinical observation, laboratory tests, and preventive strategies, which tend to be reactive, labor-intensive, and constrained by delays in identification and reaction.

In recent years, Artificial Intelligence (AI) and Machine Learning (ML) have become crucial instruments in tackling the complex difficulties presented by EUS. Utilizing extensive data from environmental monitoring systems, fish health evaluations, and genetics, AI and ML provide swift diagnosis, predictive modeling, and tailored intervention tactics. These technologies have proven capable of early disease outbreak detection, optimizing water quality and agricultural practices, and recommending exact treatments, thereby considerably improving the efficiency and sustainability of aquaculture operations.

This chapter examines the utilization of AI and ML in the diagnosis, prevention, and management of EUS. It examines AI-driven diagnostic methodologies, encompassing image-based lesion identification and predictive analytics, which improve precision and diminish reaction durations. Additionally, it analyzes how machine learning algorithms predict epidemics by incorporating environmental and historical data, thereby informing preventive strategies. The significance of AI in enhancing treatment procedures and creating novel monitoring systems is emphasized.

The chapter finishes by discussing the problems associated with the implementation of AI and ML technologies, including data restrictions, adoption

Podeti Koteshwar Rao

barriers, and technological constraints, while offering future perspectives and ideas to promote collaboration and creativity. This investigation highlights the revolutionary capacity of AI and ML in addressing EUS concerns and promoting aquaculture sustainability.

What is Endoscopic Ultrasound (EUS)?

Epizootic Ulcerative Syndrome (EUS) is a severe infectious disease impacting freshwater and brackish-water fish, primarily caused by the fungal pathogen Aphanomyces invadans. It is a complex illness, frequently aggravated by secondary bacterial and viral illnesses. EUS is marked by the formation of necrotic ulcers on the skin, resulting in considerable tissue damage, diminished physiological function, and elevated death rates. Initially recognized in the 1970s, the illness has proliferated throughout Asia, Africa, and certain regions of the Americas, emerging as one of the most commercially and ecologically detrimental diseases in aquaculture.

Symptoms of EUS manifest as tiny hemorrhage lesions on the skin that advance to extensive ulcerative wounds. These sores frequently reveal the underlying musculature and may lead to secondary infections, thereby exacerbating the health of the affected fish. The illness affects numerous species, including commercially significant fish such as catfish, tilapia, and carp, rendering it a global issue for aquaculture enterprises.

Economic and Ecological Importance

The economic influence of EUS on aquaculture is significant. The disease causes substantial production losses due to elevated mortality rates and diminished market value of fish exhibiting obvious lesions. EUS outbreaks can result in severe financial repercussions for small-scale fish farmers, who are essential to aquaculture in numerous developing areas, hence intensifying poverty and food insecurity.

EUS outbreaks ecologically damage aquatic habitats by impacting both cultivated and wild fish populations. The illness can be transmitted from aquaculture systems to natural water bodies, jeopardizing biodiversity and disrupting ecological balances. Initiatives to treat the disease, including the application of pesticides and antibiotics, may adversely affect the environment, underscoring the necessity for sustainable management techniques.

CHALLENGES IN ENDOSCOPIC ULTRASOUND MANAGEMENT

Constraints of Conventional Diagnostic Methods

Traditional diagnostic techniques for EUS, including histopathology, microbiological cultures, and PCR assays, are frequently laborious, resource-demanding, and reliant on laboratory infrastructure. Although these approaches are dependable, they are not consistently available to small-scale farmers, particularly in isolated or resource-constrained environments. The interval between sample collection, processing, and result acquisition frequently facilitates further disease dissemination, hence diminishing the efficacy of control efforts.

Control Techniques and Their Constraints

Preventive strategies, including water quality control, immunization, and antibiotic administration, have had little efficacy owing to the intricate nature of the disease and the absence of early detection methods. The overuse of antibiotics has heightened worries regarding antimicrobial resistance, hence hindering disease control initiatives.

The Necessity for Expeditious, Precise, and Scalable Solutions

To tackle these difficulties, there is an urgent requirement for technologies that can deliver rapid, precise, and economical illness detection, together with predictive capabilities to facilitate prompt therapies. Advanced technologies must be scalable to accommodate the varied requirements of global aquaculture systems.

THE FUNCTION OF ARTIFICIAL INTELLIGENCE AND MACHINE LEARNING IN DISEASE MANAGEMENT

Fundamentals of Artificial Intelligence and Machine Learning Concepts

Artificial Intelligence (AI) denotes the emulation of human cognitive functions in machines, allowing them to execute activities including learning, thinking, and decision-making. Machine Learning (ML), a branch of AI, encompasses algorithms that acquire knowledge from data to generate predictions or judgments autonomously, without explicit programming. Machine learning methodologies, including supervised learning, unsupervised learning, and deep learning, are very effective for the analysis of intricate datasets.

Addressing Intricate Challenges in Aquaculture

Artificial Intelligence and Machine Learning provide transformational capabilities for tackling the issues associated with Endoscopic Ultrasound management. These tools can evaluate extensive datasets from various sources, including environmental monitoring systems, fish health evaluations, and genomic research, to deliver actionable insights. AI-driven diagnostic instruments can accurately detect EUS-associated lesions through image recognition algorithms, whilst ML models can forecast disease outbreaks by analyzing environmental factors and historical data.

AI and ML can substantially improve the efficiency and sustainability of aquaculture operations through real-time disease surveillance, predictive modeling, and optimum intervention tactics. Moreover, these technologies may connect resource-abundant and resource-constrained environments, offering scalable solutions advantageous to both small-scale farmers and industrial aquaculture.

This chapter examines the capacity of AI and ML to transform EUS management, emphasizing their roles in diagnosis, prevention, and control. This review underscores the essential significance of these technologies in promoting the sustainability and resilience of global aquaculture systems, addressing both advancements and difficulties.

DIAGNOSTIC METHODOLOGIES FOR ENDOSCOPIC ULTRASOUND UTILIZING ARTIFICIAL INTELLIGENCE/MACHINE LEARNING

Data Acquisition and Preparation

Efficient AI/ML diagnostic methods for Epizootic Ulcerative Syndrome (EUS) depend on the quality and thoroughness of the collected data. Precise and implementable insights necessitate the amalgamation of many data sources to establish a solid basis for algorithmic evaluation.

Significance of Water Quality Metrics, Indicators of Fish Health, and Pathogen Information

Environmental variables, including temperature, pH, dissolved oxygen levels, salinity, and ammonia concentrations, are essential in assessing fish health and the probability of disease outbreaks. Monitoring these characteristics enables predictive models to associate environmental changes with the incidence of EUS. Moreover, visual indications of piscine health, such as atypical swimming patterns, alterations in pigmentation, and the manifestation of dermal lesions, offer critical diagnostic insights. Data unique to pathogens, such as fungal, bacterial, or viral strains identified during outbreaks, are essential for enhancing diagnostic specificity and comprehending disease progression.

DATA SOURCES

Sensors and Internet of Things Devices

Aquaculture farms progressively utilize IoT-enabled sensors to monitor water quality metrics in real time. These devices deliver continuous data streams, enabling ML models to identify anomalies from standard settings.

Visual Systems

High-resolution cameras and underwater drones document photographs and videos of fish populations. These visual datasets are essential for training computer vision algorithms to recognize EUS-related lesions and evaluate fish behavior.

Laboratory Findings

Histopathology data, PCR assays, and microbial cultures augment field data by verifying pathogen presence and disease stages, thereby enhancing the training datasets for AI systems.

Procedures for Data Preprocessing

Data Cleaning and Normalization: Guarantees the elimination of noise, absent values, and discrepancies from datasets.

Feature Engineering: Extracts pertinent information such as lesion dimensions, water quality indicators, and pathogen signatures to enhance model efficacy.

Data Augmentation: Expands datasets by changes (*e.g.*, image rotation or cropping) to rectify class imbalances and improve generalizability.

ARTIFICIAL INTELLIGENCE/MACHINE LEARNING MODELS FOR DIAGNOSIS

Identification Image Recognition for Ulcer

Convolutional Neural Networks (CNNs), a category of deep learning algorithms, are exceptionally proficient in processing picture data. CNN-based systems are capable of:

Identify and categorize EUS-related lesions with high precision through the analysis of visual patterns.

Distinguish between healthy fish and those impacted by EUS or other ailments.

Models trained on annotated datasets of fish photos have demonstrated exceptional accuracy in detecting early-stage ulcers, facilitating prompt therapies.

Predictive Models for Early Detection of Outbreaks

Supervised machine learning techniques, such as Random Forests, Gradient Boosting Machines, and Deep Neural Networks, are utilized to forecast EUS epidemics. These models examine historical data, encompassing environmental conditions and previous outbreaks, to discern patterns and predict disease onset. Time-series analysis and Recurrent Neural Networks (RNNs) are capable of processing temporal data, hence enhancing prediction accuracy.

The Function of Natural Language Processing (NLP) in Literary Analysis

NLP methodologies, including subject modeling and named entity identification, are employed to analyze enormous volumes of scholarly literature on EUS. NLP facilitates the extraction of pertinent ideas from research papers, technical reports, and agricultural records by:

Recognizing knowledge deficiencies and nascent trends.

Proposing diagnostic and therapeutic regimens grounded in current research.

Case Analyses and Implementations

PRACTICAL ILLUSTRATIONS OF AI-ENHANCED DIAGNOSTIC INSTRUMENTS

Platforms for Monitoring Fish Health

AI-driven platforms incorporate picture recognition, water quality sensors, and machine learning algorithms to deliver real-time health evaluations. A technique implemented in Southeast Asia precisely diagnosed EUS lesions in catfish with an accuracy exceeding 95%, thereby reducing diagnostic time by 70%.

Predictive Analytics for Epidemic Prevention:

A pilot initiative in India utilized machine learning models to forecast EUS outbreaks in carp farms by analyzing meteorological and water quality data, with an 80% accuracy rate in predicting high-risk intervals.

Comparisons with Conventional Methods

Aspect Conventional Approaches Artificial Intelligence Techniques

Precision

Dependent on manual observation, susceptible to inaccuracies.

Elevated precision attributable to sophisticated pattern recognition.

Velocity

Laborious owing to sample processing.

Immediate or near-instantaneous analysis.

Scalability

Restricted to laboratories with appropriate equipment.

Applicable in various agricultural environments.

Significant operational expenses.

Long-term cost-effectiveness following deployment.

These comparisons underscore the transformative capacity of AI and ML in surmounting the constraints of conventional diagnostic methods, facilitating more efficient and sustainable EUS management.

This section highlights how AI and ML improve the precision and efficiency of EUS diagnosis while also broadening access to advanced diagnostic tools, helping both small-scale farmers and large aquaculture businesses.

AI/ML-SUPPORTED PREVENTION STRATEGIES

Predictive Modeling

Predictive modeling is fundamental in averting Epizootic Ulcerative Syndrome (EUS) epidemics by detecting trends and dangers before their occurrence. Machine Learning (ML) systems utilize historical and real-time data to predict the probability of outbreaks, facilitating proactive measures.

Machine Learning Models for Predicting Outbreaks

Machine learning methods, including Random Forests, Gradient Boosting Machines, and Neural Networks, evaluate intricate, multidimensional datasets comprising water quality metrics, fish health indicators, and historical disease occurrences. These algorithms can forecast prospective outbreaks by detecting links between environmental factors and the emergence of EUS.

For Example

Random Forests: Efficient in managing datasets with missing values and non-linear correlations, delivering reliable forecasts even in turbulent conditions.

Deep Learning Models: Convolutional Neural Networks and Recurrent Neural Networks are proficient at interpreting spatial and temporal data, respectively, hence improving the precision of epidemic predictions.

Convergence of Meteorological and Hydrological Models

Climatic and hydrological factors substantially affect the incidence of EUS. The integration of meteorological models with aquaculture data yields significant insights into disease patterns. For instance:

Increased water temperatures or atypical rainfall patterns can facilitate the spread of *Aphanomyces invadans.*

Hydrological models evaluate water flow and connectivity between aquaculture systems and natural water bodies, forecasting potential areas of spread.

Integrated frameworks allow farms to implement region-specific preventive strategies adapted to environmental conditions.

Enhancement of Procedures

AI-driven optimization guarantees sustainable aquaculture practices by harmonizing production efficiency with disease mitigation.

Artificial Intelligence for Water Quality Management

Ensuring adequate water quality is crucial for preventing EUS outbreaks. Artificial intelligence systems analyze real-time data from Internet of Things sensors, delivering actionable insights for sustaining optimal circumstances. Illustrations encompass:

Automated notifications for pH discrepancies, diminished oxygen levels, or increased ammonia concentrations.

Predictive algorithms suggest modifications to aeration, filtration, or water changes according to anticipated conditions.

Feeding Regimens and Behavioral Observation

Inadequate feeding methods can impair fish immune systems, heightening vulnerability to EUS. AI-driven feeding systems assess fish behavior and appetite, optimizing feed amounts and timing to reduce waste and enhance health. AI systems also identify atypical swimming behaviors or lethargy, early signs of possible illness.

Instruments for Evaluating Disease Risk

Machine learning-based risk assessment algorithms evaluate the probability of disease outbreaks by taking into account:

Environmental characteristics peculiar to the farm.

Historical epidemiological data and management methodologies.

These tools assist farmers in prioritizing preventive steps, such as enhancing biosecurity or scheduling prophylactic treatments, according to the assessed risk levels.

Decision Support Systems

AI-driven Decision Support Systems (DSS) enable aquaculture managers and farmers by delivering accessible, actionable insights *via* intuitive interfaces.

Creation of Intuitive Platforms

AI-driven Decision Support Systems amalgamate data from sensors, predictive models, and external sources (such as meteorological forecasts) into a cohesive dashboard. The characteristics of these platforms encompass:

Real-Time Monitoring: Presenting water quality indicators, ichthyological behavior, and system integrity at a glance.

Predictive Alerts: Notifications for elevated-risk conditions, facilitating prompt preventive measures.

Interactive Recommendations: Methodical instructions for mitigating potential risks, customized to particular agricultural conditions.

These systems are frequently engineered with multilingual capabilities and user-friendly interfaces to accommodate various user demographics, including small-scale farmers with less technological proficiency.

Mobile Applications and Remote Access

Numerous DSS platforms are now available through mobile applications, guaranteeing continuous connectivity and oversight. Remote access functionalities enable farmers to:

Supervise their aquaculture systems remotely.

Obtain automated reports that encapsulate essential health and environmental parameters.

Cooperation and Information Exchange

Advanced DSS platforms enhance collaboration by permitting data sharing among stakeholders. For example:

Researchers can examine aggregated data to reveal novel insights into EUS.

Agricultural producers can evaluate their methods against industry benchmarks.

Practical Illustrations

Epi-Aqua DSS: A platform integrating water quality sensors, disease risk models, and AI-generated suggestions to support farmers in Southeast Asia. This approach markedly decreased illness incidence in participating farms by 30%.

SmartAquaculture.io: A smartphone application providing predictive analytics and decision-making resources for aquaculture health management, available to small-scale producers in remote areas.

Control Techniques Augmented by Artificial Intelligence/Machine Learning

Systems for Recommending Treatments

AI and ML technologies markedly improve the accuracy and effectiveness of treatment protocols for Epizootic Ulcerative Syndrome (EUS), transitioning from broad methods to individualized disease management.

ARTIFICIAL INTELLIGENCE INSTRUMENTS FOR RECOMMENDING ANTIBIOTICS OR ALTERNATIVE THERAPIES

Analysis of Resistance Patterns

AI systems examine large databases regarding antibiotic utilization and disease resistance trends, facilitating accurate treatment recommendations. For instance:

If Aphanomyces invadans is linked to secondary bacterial illnesses, AI systems can determine the most efficacious medications while circumventing those with elevated resistance rates.

AI can propose environmentally sustainable options, such as probiotics or herbal remedies, tailored to individual agricultural conditions.

Mitigating Antibiotic Overutilization

By endorsing specific treatments, AI reduces the excessive use of antibiotics, hence alleviating issues linked to antimicrobial resistance and environmental pollution.

Customized Therapeutic Strategies Utilizing Machine Learning Algorithms

Machine Learning (ML) models evaluate several characteristics, including fish species, age, health state, and environmental circumstances, to formulate customized treatment programs. These tailored strategies enhance therapy efficacy and minimize superfluous interventions.

An ML model, utilizing past treatment outcomes, forecasts the ideal combination of antifungal drugs and water quality modifications for a particular outbreak scenario, facilitating effective recovery while minimizing fish stress.

Surveillance Systems

Ongoing disease surveillance is essential for prompt intervention and efficient management of EUS. IoT and AI-driven solutions provide real-time monitoring, enhancing agricultural resilience to disease outbreaks.

INTERNET OF THINGS AND ARTIFICIAL INTELLIGENCE FOR ONGOING DISEASE MONITORING

Sensor Networks for Monitoring Water Quality

IoT-enabled sensors installed in aquaculture ponds gather data on pH, dissolved oxygen, ammonia, and temperature. AI systems analyze this data instantaneously to detect anomalies associated with EUS outbreaks.

Systems for Monitoring Behavior

Underwater cameras and drones with AI algorithms observe fish behavior. Unusual behaviors, such as lethargy, irregular swimming, or surface gasping, activate early alerts, enabling farmers to inspect and respond swiftly.

Convergence of Environmental and Biological Data

Advanced methods amalgamate environmental monitoring with biological markers, like growth rates and lesion development, to furnish a comprehensive perspective on fish health.

EARLY WARNING SYSTEMS ACTIVATED BY ANOMALIES

Alerts Based on Thresholds

AI systems determine standard ranges for environmental and behavioral parameters. Exceeding certain thresholds automatically activates alerts, necessitating precautionary measures.

Predictive Notifications Utilizing Machine Learning

Machine learning methods examine previous outbreak data to predict hazards before aberrations, providing a crucial opportunity for intervention.

A monitoring system identifies a decrease in dissolved oxygen levels and slight behavioral alterations. AI models forecast a significant likelihood of EUS commencement within days, allowing farmers to modify aeration systems and implement preventive water treatments.

Integration of Genomics and Bioinformatics

The integration of AI with genomics and bioinformatics creates new opportunities for regulating EUS at the genetic level.

EMPLOYING AI TO DETECT RESISTANT FISH STRAINS

Analysis of Genomic Data

Artificial intelligence algorithms examine genetic data to pinpoint markers linked to disease resistance in fish populations. This information directs selective breeding initiatives to improve the robustness of aquaculture species.

Models for Predictive Breeding

Machine learning techniques forecast child performance based on parental genetics, expediting the creation of resistant breeds.

Genomic investigations of carp identify particular alleles associated with decreased sensitivity to Aphanomyces invadans. AI systems emphasize these alleles in breeding programs, yielding generations of more resilient fish.

CRISPR AND GENE-EDITING TECHNIQUES INFORMED BY MACHINE LEARNING

Artificial Intelligence in Target Identification

AI systems discern prospective gene-editing targets by examining pathogen-host interactions and genetic susceptibilities in fish.

Enhancing CRISPR Approaches

Machine learning models predict the results of several gene-editing approaches, enhancing CRISPR methods for the creation of disease-resistant fish.

Researchers utilize AI to identify immune-response genes in tilapia and employ CRISPR to augment their antifungal characteristics, resulting in fish that are less susceptible to EUS, even in inadequate environments.

PRACTICAL APPLICATIONS

Artificial Intelligence-Enhanced Health Platforms

Platforms such as Fish Health AI incorporate treatment recommendation algorithms, monitoring instruments, and genetic analysis, providing holistic disease management solutions.

Success Stories in Gene Editing

AI-directed CRISPR initiatives in aquaculture have effectively generated shrimp that are immune to viral pathogens, illustrating the viability of employing analogous methodologies for EUS.

Obstacles and Prospective Developments

Although AI and ML provide transformative potential in EUS control, difficulties, including data availability, integration expenses, and the necessity for user-friendly interfaces, persist. Overcoming these obstacles necessitates:

Cooperation among technologists, aquaculture specialists, and geneticists.

Enhanced utilization of open-data methodologies for improved AI model training.

Government and business assistance to subsidize the installation of technology for small-scale farms.

Utilizing AI/ML for tailored therapy, ongoing surveillance, and genomic advancements, the aquaculture sector can markedly enhance its ability to manage EUS, promoting healthier fish populations and more sustainable practices.

Obstacles and Constraints

Despite the revolutionary promise of AI and ML in the management of Epizootic Ulcerative Syndrome (EUS), numerous hurdles and constraints impede their general acceptance and efficacy in aquaculture techniques.

CONSTRAINTS OF DATA

Concerns Regarding Data Quality

The efficacy of AI/ML models is significantly contingent upon the quality of the data utilized for training. Nevertheless, aquaculture data frequently experience:

Noise and Inconsistencies: Variability in sensor data, human observational errors, and poor documentation can result in erroneous predictions.

Dataset Bias: Data largely sourced from particular regions or species may lack generalizability across varied aquaculture contexts, hence constraining model applicability.

Data Accessibility

Numerous small-scale farms lack comprehensive data-gathering methods, resulting in insufficient datasets for AI/ML training. In developing nations, inadequate access to dependable internet connectivity exacerbates the challenges of real-time data exchange and integration.

Absence of Standardization

The lack of established standards for data collection, storage, and sharing within the aquaculture business presents a considerable difficulty. Divergences in data formats and quality measures hinder model building and restrict interoperability among systems.

TECHNOLOGICAL OBSTACLES

Requirement for Computational Resources and Expertise

AI/ML applications sometimes necessitate considerable computational resources for model training, especially when utilizing deep learning or extensive genomic datasets. Nonetheless:

Numerous aquaculture farms, particularly in low-income areas, are devoid of high-performance computing technology.

The development and deployment of AI solutions necessitate proficiency in data science and aquaculture, resulting in a skills gap that constrains implementation.

Model Interpretability and Reliability

The intricacy of numerous AI models, particularly deep learning algorithms, renders them challenging to interpret. For example:

A lack of clarity on the rationale behind predictions, such as suggested treatments, may hinder farmers' faith in model outcomes.

Regulatory authorities may mandate explainable models to substantiate their application in crucial choices such as antibiotic delivery or genetic modification.

Challenges of Integration

Integrating AI/ML technologies with current aquaculture systems can be technologically challenging, especially when merging IoT devices, environmental monitoring systems, and genomic databases.

OBSTACLES TO ADOPTION

Expenditures

The substantial initial capital expenditure necessary for implementing AI-driven systems, encompassing IoT devices, software platforms, and staff training, poses a considerable obstacle for several small-scale farms. Despite the potential for long-term cost benefits, the initial expenditures hinder implementation.

Insufficient Awareness

A multitude of stakeholders within the aquaculture sector lack familiarity with AI and ML technologies, their functionalities, and their prospective advantages. This deficiency in awareness engenders reluctance in embracing novel solutions, especially among older or less technologically adept farmers.

Opposition to Technological Integration

Cultural and behavioral opposition to change constitutes another substantial obstacle. For instance:

Farmers familiar with conventional techniques may perceive AI/ML technologies as excessively intricate or superfluous.

Apprehensions over data privacy and ownership may deter participation, particularly when external platforms gather and scrutinize agricultural data.

TACKLING THE CHALLENGES

Enhancing Data Infrastructure

Implementing uniform data collection and dissemination methods throughout the industry.

Advocating for the utilization of open data sources to improve dataset diversity and model resilience.

Improving Accessibility and Education

Governments and industry stakeholders might subsidize technological expenses and offer incentives for small-scale farmers to use AI tools.

Training programs centered on the practical uses of AI/ML in aquaculture can address the skills deficit.

Concentrating on Explainable Artificial Intelligence

Creating interpretable models, such as decision trees or rule-based systems, whenever feasible.

Integrating intuitive graphics to assist farmers in comprehending forecasts and advice.

Establishing Trust and Recognition

Exhibiting success narratives and case studies to illustrate the concrete advantages of AI in enhancing disease management and agricultural productivity.

Involving farmers in the development process to address their issues and customize solutions to their requirements.

By tackling these obstacles, the aquaculture sector may realize the complete potential of AI and ML, facilitating more efficient, sustainable, and scalable solutions to handle EUS and other aquaculture issues.

Obstacles and Constraints

Despite the considerable promise of AI and ML in combating Epizootic Ulcerative Syndrome (EUS) in aquaculture, numerous hurdles and constraints hinder their effective implementation and efficacy. These can be classified as data-related challenges, technical impediments, and adoption constraints.

CONSTRAINTS OF DATA

Concerns Regarding Data Quality

The efficacy of AI/ML models depends on superior data quality. Regrettably, data gathered in aquaculture environments frequently encounter:

Noise and Errors: Variability in sensor measurements, inaccuracies in manual data entry, and inadequate datasets might result in incorrect model results.

Insufficient Data Labeling: Inconsistent labeling of ichthyopathologies or environmental variables undermines supervised learning algorithms that depend on precisely labeled training data.

Data Accessibility

Small-scale and resource-constrained aquaculture enterprises often lack the infrastructure necessary for systematic and thorough data collection. Challenges encompass:

Limited Access to IoT Devices and Data Logging Systems

Insufficient historical data regarding disease outbreaks and agricultural circumstances, especially in developing areas.

Absence of Standardization

No globally accepted standards exist for data collecting, formatting, or storage in aquaculture. This presents issues in:

Consolidating datasets from many sources.

Creating generalizable AI/ML models applicable to many aquaculture systems.

TECHNICAL DIFFICULTIES

Requirement for Computational Resources and Expertise

Artificial Intelligence and Machine Learning models, particularly those utilizing deep learning, necessitate substantial computer resources for training and real-time implementation. Nevertheless:

Numerous Aquaculture Facilities are Devoid of High-Performance Computing Tools

The aquaculture sector lacks expertise in data science and AI, resulting in a skills gap for developing and managing these technologies.

Model Interpretability and Reliability

Advanced AI/ML models, such as deep neural networks, frequently act as "black boxes," making it impossible to explain their decision-making processes. This generates:

Trust Issues: Farmers and managers may hesitate to depend on model outcomes if they lack comprehension of the underlying rationale.

Regulatory Challenges: Government entities and certification organizations may want explicit rationales for AI-recommended interventions, particularly regarding medicines or biosecurity protocols.

Integration with Current Systems

Numerous farms utilize conventional systems and processes. Integrating AI tools frequently necessitates substantial modifications, encompassing:

Enhancing IoT-integrated infrastructure

Instructing staff on the utilization of digital dashboards and the interpretation of model forecasts.

OBSTACLES TO ADOPTION

Expenditures

The deployment of AI/ML systems necessitates substantial investment in:

Internet of Things devices and sensors for data acquisition.

Software platforms for data processing and visualization.

Instruction and assistance for end-users.

Small-scale farmers, constituting the majority of the aquaculture sector, frequently lack the financial capacity to cover these initial expenses, notwithstanding the prospective long-term advantages.

Insufficient Awareness

A significant number of aquaculture players lack familiarity with AI and ML technology.

Doubt regarding the tangible advantages of AI in disease management.

Fallacies suggesting that AI tools are excessively intricate or superfluous for small-scale enterprises.

Opposition to Technological Integration

Resistance to new technologies, both behavioral and cultural, is prevalent, especially in conventional agricultural areas. Issues encompass:

Apprehension Regarding Employment Termination: The belief that automation could diminish the necessity for the human workforce.

Concerns Regarding Data Privacy: Hesitance to disclose farm-specific data to external platforms or centralized databases.

CONFRONTING OBSTACLES

Enhancing Data Infrastructure

Implementing standardized processes for data collection and reporting in aquaculture.

Advocating for the utilization of open-access datasets, fostering collaboration, and enhancing model generalization.

Minimizing Expenditures and Improving Accessibility

Governments and groups may offer subsidies or affordable technology packages specifically designed for small-scale farmers.

Creating affordable AI platforms with minimal hardware specifications can eliminate financial obstacles.

DEVELOPING PROFICIENCY AND CONSCIOUSNESS

Implementing training programs for farmers to elucidate AI and ML technology.

Presenting effective case studies to demonstrate the concrete advantages of implementing AI in illness management.

Concentrating on Explainable Artificial Intelligence

Creating more straightforward, rule-based models or hybrid methodologies that integrate explainability with efficacy.

Integrating accessible graphics and intuitive interfaces to bolster farmer trust and involvement.

By tackling these difficulties, the aquaculture sector may harness the capabilities of AI and ML for EUS management, facilitating healthier fish populations, sustainable practices, and enhanced economic results.

PROSPECTIVE INSIGHTS AND SUGGESTIONS

The management of Epizootic Ulcerative Syndrome (EUS) and other aquaculture difficulties will depend on the ongoing advancement and implementation of AI and ML technologies. Innovative trends and collaborative methods can revolutionize the aquaculture sector, making it more sustainable and resilient.

Emerging Trends in Artificial Intelligence and Machine Learning within Aquaculture

Progress in AI and ML technologies is creating new prospects for aquaculture. Principal trends encompass:

Sophisticated Predictive Analytics

Machine learning algorithms that can use multi-modal data, including water quality, fish behavior, and environmental variables, to enhance the accuracy of disease outbreak predictions.

Real-time model adaptation employing continuous learning methodologies to enhance predictions utilizing new data.

Generative Artificial Intelligence in Bioinformatics

AI-facilitated identification of novel antimicrobial peptides or probiotics specifically designed to address *Aphanomyces invadans*.

Advanced genome-editing techniques utilizing AI to pinpoint and optimize genes linked to disease resistance.

Automated Aquaculture Systems

AI-powered autonomous underwater drones and robotic systems for ongoing health monitoring, feeding, and water management.

Blockchain Integration for Traceability and Transparency

The amalgamation of AI/ML with blockchain technology presents unparalleled advantages for aquaculture:

End-to-End Traceability: Blockchain guarantees safe, immutable documentation of fish health, water quality, and treatment history, fostering confidence among stakeholders and customers.

Augmented Transparency: Integrating AI insights with blockchain technology can produce authenticated datasets for regulatory adherence, sustainable certification, and market entry. The block diagram of the AI-based EUS Detection System is shown in Fig. (**1**).

Cooperative Endeavors among Stakeholders

Collaboration among industry players, such as farmers, researchers, technology developers, and politicians, is essential for surmounting obstacles and enhancing AI/ML adoption.

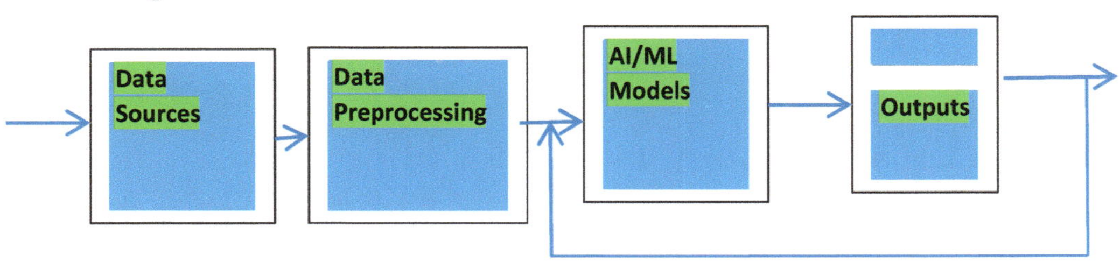

Fig. (1). Block Diagram of the AI-Based EUS Detection System.

Fraud Prevention: Blockchain technology can avert data manipulation and guarantee the legitimacy of AI-generated insights, hence enhancing credibility.

A blockchain system combined with an AI-driven illness monitoring software can deliver verified alarms and treatment records, thereby ensuring responsibility and fostering trust in supply chains.

Data Sharing Initiatives

Creating centralized or federated databases for the dissemination of anonymised aquaculture data can improve the training and applicability of AI models across various geographies.

Establishing universal standards for data gathering and reporting will facilitate the smooth integration of varied information.

Collaborations Between the Public and Private Sectors

Governments and corporate organizations can together invest in the creation of accessible AI technologies and infrastructure designed for small-scale farmers.

Joint funding for research and development can expedite innovation in AI-based disease management.

Capacity Building: Educational seminars and workshops for agricultural producers and aquaculture administrators on the practical application of AI technologies. Creation of multilingual, culturally pertinent AI platforms to guarantee inclusion.

ILLUSTRATION: SCHEMATIC REPRESENTATION OF THE AI-DRIVEN EUS DETECTION SYSTEM

Regulation of EUS

With studies on EUS having definitively established certain causal elements, evidence-based control methods can be formulated and executed.

Prevention

Considering that EUS frequently manifests and is typically most detrimental in wild fish populations, managing outbreaks within a localized region can be exceedingly challenging. Consequently, in regions where EUS is not endemic, the most efficacious method of control would be to avert the introduction of the illness into the country or a designated area within the country (*e.g.*, an EUS-free island) initially. Identified EUS-sensitive fish are prevalent in Africa and Central Asia, whereas possibly susceptible species are found in numerous other nations now free from EUS. Proposed quarantine and health certification protocols for the translocation of live fish between countries or regions may effectively mitigate the dissemination of EUS to new territories (Humphrey *et al.* 1997). The dissemination of standardized diagnostic methods for significant fish diseases, such as EUS, along with the establishment of pathogen-host databases, will enhance the application of these guidelines.

In regions where EUS is currently deemed endemic, preventative initiatives should encompass the following activities:

- Elimination
- Exclusion
- Administration
- Monitoring
- Intervention and Elimination

The objective is to eliminate *A. invaders* from a previously contaminated location (e.g., agricultural site or aquatic environment). While little information exists regarding the fungus's survival between outbreaks, current evidence indicates that infection may endure, typically at low prevalence, within susceptible fish populations present at the site between crops. It is also feasible, but less probable, that the fungus may persist in the aquatic environment either as encysted spores or on non-fish substrates.

Consequently, strategies to eliminate the fungus should encompass:

- Elimination of all fish, especially all vulnerable species, from ponds, reservoirs, and water supply channels before re-stocking;
- Desiccation and liming of ponds;
- Sanitization of tainted apparatus.

Exclusion

After the fungus has been eliminated from a location, it is crucial to avert reintroduction. Infection is likely transmitted by infected or carrier fish, as well as through contaminated water or equipment. Consequently, the subsequent steps should be evaluated, considering local conditions and the anticipated prevalence of EUS in wild populations.

Seed stock and brood stock should, if feasible, be sourced from EUS-free sites and prophylactically treated for exterior fungal infections (*e.g.*, using a 1% NaCl bath treatment) before being introduced to the site.

All wild fish must be strictly prohibited from farms in endemic regions

Water should, if feasible, be sourced only from an EUS-free origin, such as a well or borehole. Should water from a potentially contaminated source be utilized, it must be filtered through fine screens at the supply inlet (to reduce the risk of wild fish ingress) and retained in a fish-free reservoir for a minimum of 10 days (Mathews and Reynolds, 1990, demonstrated that A. astaci spores remained viable for 6-9 days at 10°C, and a comparable duration has been established for A. invadans (Campbell, unpublished)). If the risk of introducing contaminated wild fish or A. When the risk of invasive propagules is deemed significant (*e.g.*, during an ongoing EUS outbreak in wild fish), it is imperative to treat the stored water with piscicides and a disinfectant before utilizing it in ponds.

Equipment potentially utilized at infected locations must be disinfected using normal hypochlorite or iodophor treatments.

Administration

Epidemiological evidence indicates that EUS epidemics in aquaculture are exacerbated by elevated stocking levels. During high-risk periods, such as when EUS prevalence is elevated in neighboring wild fish populations, stocking levels should be minimized, and farmed populations should experience minimal stress. Fish must be monitored to prevent bacterial and parasitic skin pathogens from causing problems during high-risk periods, as these agents may facilitate fungal infections. Likewise, the pond ecosystem must be monitored to ensure that abiotic variables, such as low dissolved oxygen levels that could cause skin injury, remain within acceptable parameters.

An uncomplicated and efficient preventive measure, potentially suitable in certain endemic regions, is to cultivate plants that exhibit resistance to EUS. EUS has not been documented in tilapia or milkfish, and is exceedingly rare in Chinese or European carps.

REGULATION OF EUS

Monitoring

Regular monitoring of the overall health and EUS status of vulnerable fish populations during the growout period is essential. If a substantial percentage of fish exhibiting skin injury is observed during periods when A. Invadans is expected to be accessible, then the causes must be determined and suitable measures implemented. While sampling methodologies for pond fish remain inadequately delineated, it is recommended that a representative sample from each at-risk population be assessed at a minimum frequency of once per week. The optimal methods for doing this will differ from farm to farm and rely on the species cultivated. In certain instances, it may be adequate to meticulously monitor the fish during feeding, considering that a diseased segment of the population may abstain from feeding. The potential stress on the fish, especially the risk of skin damage during the collection of a representative sample, must be weighed against the need for observation.

Therapeutic Intervention

A limited number of researchers have found potentially effective strategies for limiting the transmission of EUS in farmed fish populations.

Aquarium experiments have demonstrated that certain compounds inhibit the formation of EUS lesions in abraded African catfish (*Clarias gariepinus*) fingerlings exposed to A. Invasive propagules: 25 mg/L formalin, 5 ppt sodium chloride, 5 ppm Coptrol (a chelated copper compound), and 0.1 mg/L malachite green (Callinan, unpublished).

Pond studies demonstrated that 5 ppm Coptrol inhibited the formation of EUS lesions in abraded African catfish fingerlings exposed to A. Invasive propagules. Malachite green (0.1 mg/L) demonstrated partial efficacy in inhibiting induction, but formalin (25 mg/L) was ineffectual (Callinan, unpublished).

Pond research in Bangladesh (Ahmed and Rab, 1995) indicated that the application of agricultural lime to ponds during the cultivation period mitigated the severity of EUS outbreaks. Nonetheless, in later pond experiments in the Philippines, the application of lime at 2 kg/100 m2 throughout the cultivation phase did not inhibit the production of EUS lesions in abraded African catfish (Callinan, unpublished).

In vitro studies indicate that malachite green, hydrogen peroxide, and Proxitane 0510 (which contains 5% peracetic acid in hydrogen peroxide) may exhibit effective fungicidal properties against A. invadans (Lilley and Inglis, 1997).

No published reports exist about successful curative therapies for established EUS lesions in farmed fish; however, Indian researchers assert that a patented mixture, "CIFAX," is curative.

Biological Agents Associated with EUS Infection

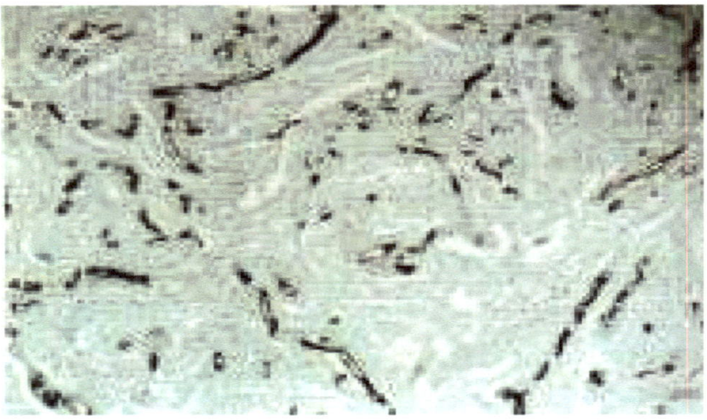

Fig. (2). *A. invadans* hyphae demonstrating the ability to grow invasively through snakehead muscle. (Grocott's stain).

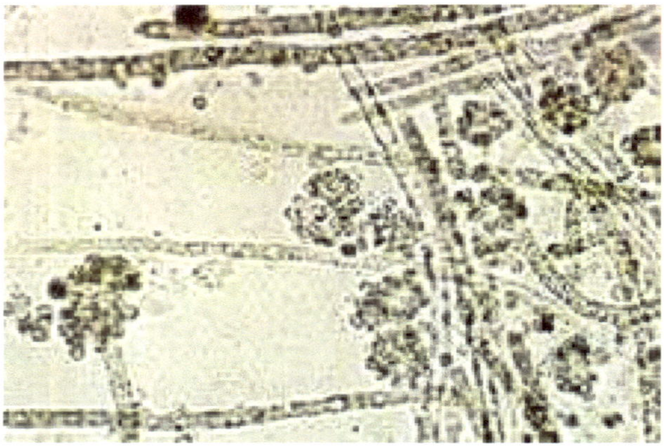

Fig. (3). Squash preparation of fungus showing typical *Aphanomyces* characteristics.

Fig. (4). Moderate dermatitis in snakehead following intramuscular injection of rhabdovirus isolates T9412. This virus is considered a probable component cause of EUS.

Fig. (5). Typical EUS-like dermal ulceration in snakehead following intramuscular injection of rhabdovirus T9412, followed by bath challenge with *Aphanomyces invadans* zoospores.

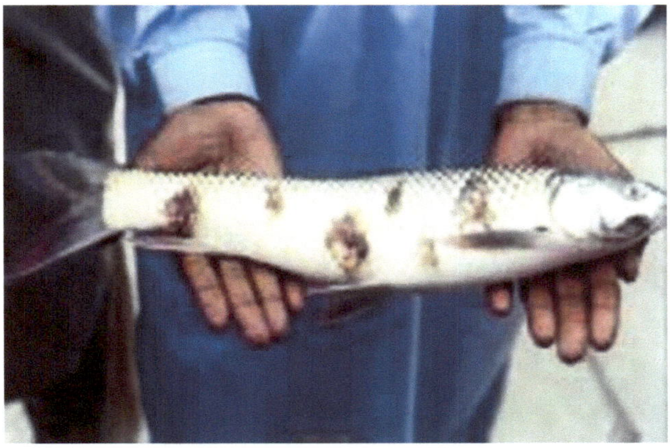

Fig. (6). Lesions caused by *Lernaea* infestation, which may predispose fish to EUS.

Fig. (7). Cutaneous *Flexibacter* bacterial infection, which may predispose fish to EUS.

Clinical Signs and Gross Pathology

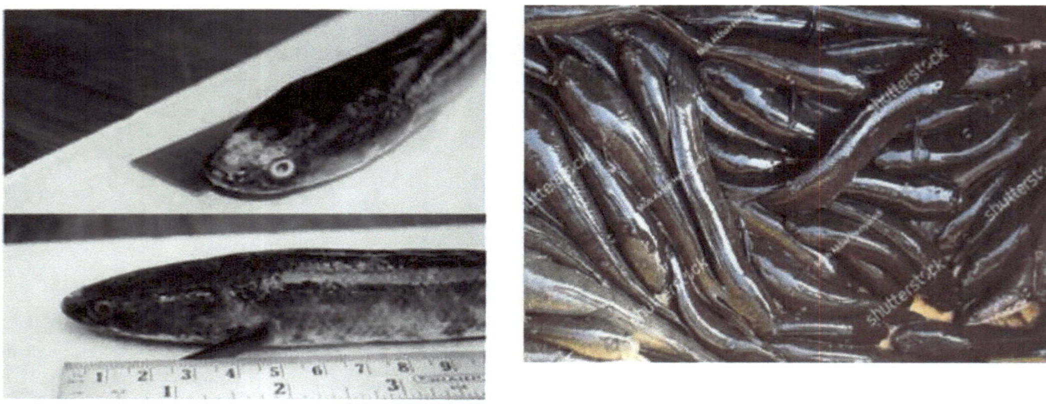

Fig. (8). Giant gouramy showing ulcers caused by a disease unrelated to EUS.
EUS snakehead fish in Warangal Local Ponds.

Fig. (9). EUS-affected snakehead, swimming with its head out of the wate.

Fig. (10). EUS-affected rohu showing moderately advanced lesions.

Biological Agents Linked to EUS Infection

Aphanomyces invadans hyphae exhibiting invasive growth within snakehead muscle tissue; Grocott's stain. Preparation of squash specimens exhibiting distinctive features of Aphanomyces fungi.

Moderate dermatitis observed in snakehead after intramuscular injection of rhabdovirus isolate T9412. This virus is regarded as a likely contributing factor to EUS. Characteristic EUS-like cutaneous ulceration in snakehead subsequent to intramuscular administration of rhabdovirus T9412 followed by immersion challenge with Aphanomyces invadans zoospores.

Lesions resulting from Lernaea infestation, which may render fish susceptible to EUS. Cutaneous Flexi bacter bacterial infection that may predispose fish to Epizootic Ulcerative Syndrome (EUS).

Clinical Manifestations and Macroscopic Pathology

Giant gourami exhibited ulcers not associated with EUS. EUS-affected snakehead fish were observed in local ponds in Warangal. A snakehead affected by EUS was seen swimming with its head above the water. Rohu was affected by EUS, exhibiting fairly advanced lesions. EUS-affected snakehead fish exhibited characteristic dermal ulceration and significant degradation of cranial tissues. *Channa striatus* affected by EUS exhibited fairly advanced skin lesions.

Histopathology

Initial infection exhibited fungal hyphae within localized regions of necrosis. Moderate EUS lesions showed mild sarcolysis. Severe mycotic granulomas were observed in a muscular lesion from an advanced case of EUS. Fungal hyphae (stained black) were seen within mycotic granulomas in a muscular lesion from an advanced case of EUS (Grocott's stain). Fungal hyphae were also present in the kidneys of fish infected with EUS, with unrelated granulomas (left and right) and mycotic granulomas (center) observed in the renal tissue. Histopathology images are presented in Figs, **1** – **11**).

Endoscopic Ultrasound in India

Channa punctatus infected by EUS was captured in Hasanparthy Lake, Warangal, Telangana, India (Photograph: PK RAO). *Channa striatus* affected by EUS was collected from the Dharmasagar Reservoir in Warangal, Telangana, India. *Channa striatus* infected by EUS was also captured in Bandham Lake, Warangal, Telangana, India.

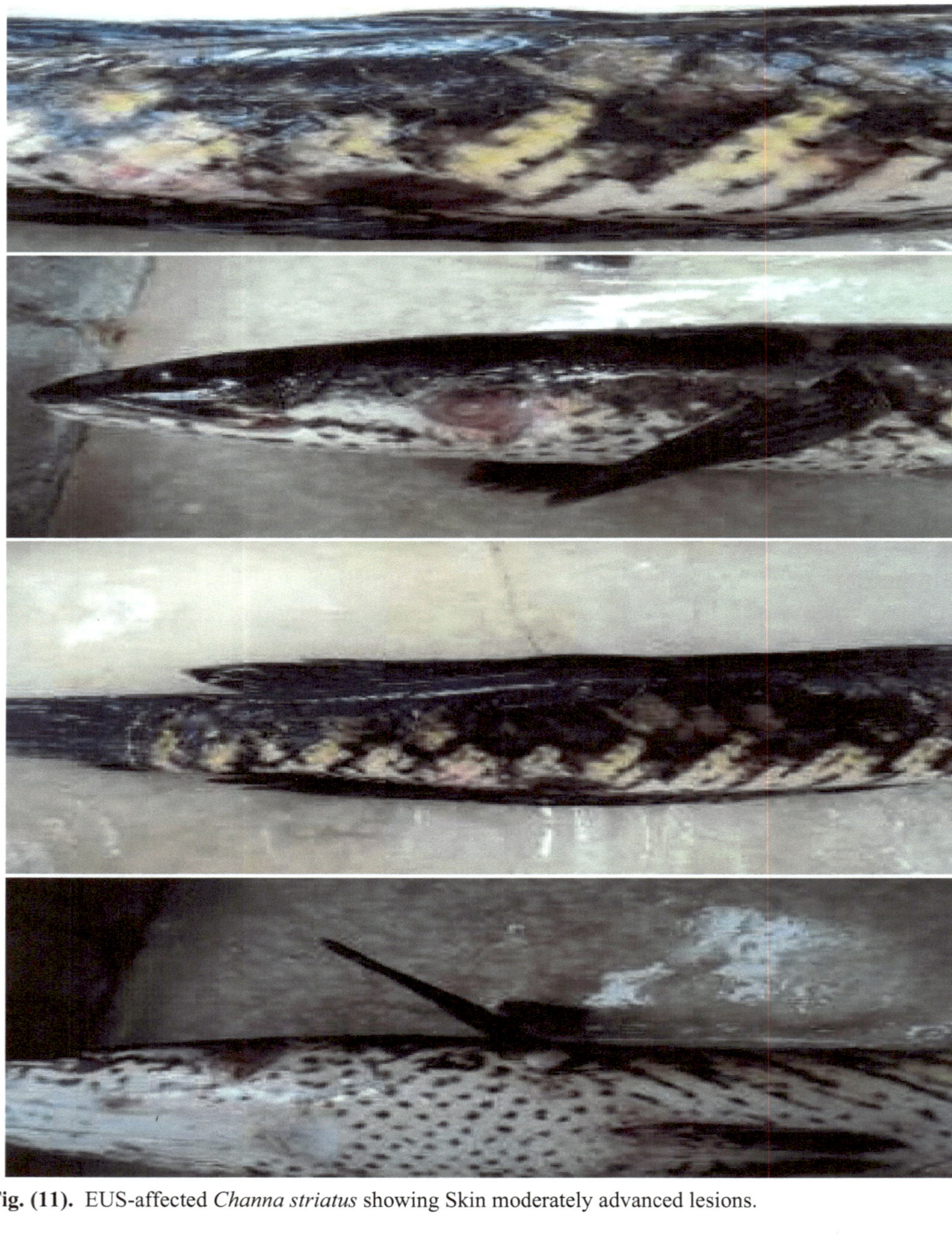

Fig. (11). EUS-affected *Channa striatus* showing Skin moderately advanced lesions.

A dermal ulcer was observed on a *Channa striatus* specimen. The pale lesions on the ulcer surface are granulomas formed in response to invasion of skeletal muscle by Aphanomyces invaders.

Examination of EUS Outbreaks

An outbreak is characterized as a temporary epidemic or a sequence of illness occurrences concentrated in a certain location and timeframe. Such disease episodes typically represent new cases of a disease manifesting at a frequency beyond the ordinary expectation. In this section, the phrases outbreak and epidemic are utilized interchangeably.

An outbreak investigation is a methodical process designed to uncover the origins and sources of epidemics, aimed at controlling the current epidemic and preventing future occurrences. The principal aim of an epidemic or disease outbreak inquiry is to ascertain methods for avoiding further spread of the pathogenic agent. The epidemiological method for outbreak investigations is founded on the principle that disease cases are not randomly dispersed, but rather manifest in patterns within the susceptible population. It is the job of the epidemiologist to document and study these patterns to help fulfill the primary purpose.

Limited information exists regarding the mechanisms by which EUS disseminates within and between regions. At the same time, the mobility of subclinically infected fish likely plays a significant role in the transmission of A. invading infection. In the absence of appropriate control and preventative measures, the illness is likely to persist in spreading and to cause recurrent outbreaks in endemic regions. Thorough research of initial outbreaks in previously EUS-free regions and recurring outbreaks in historically endemic areas is urgently required. It will provide critical insights into the causative elements of EUS. Biological agents associated with EUS infection are shown in Figs. (**12-21**).

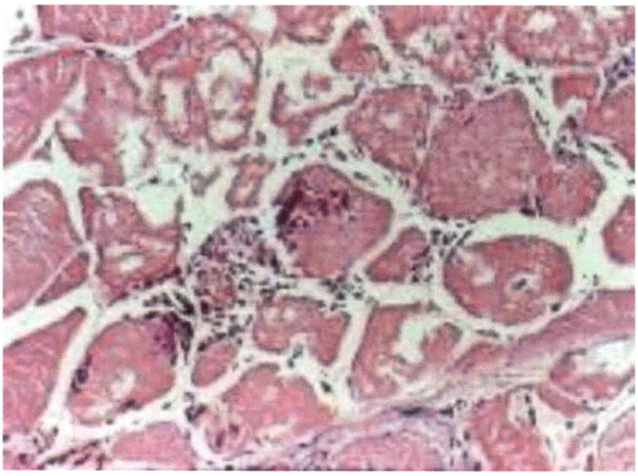

Fig. (12). Early infection demonstrating fungal hyphae within small areas of necrosis.

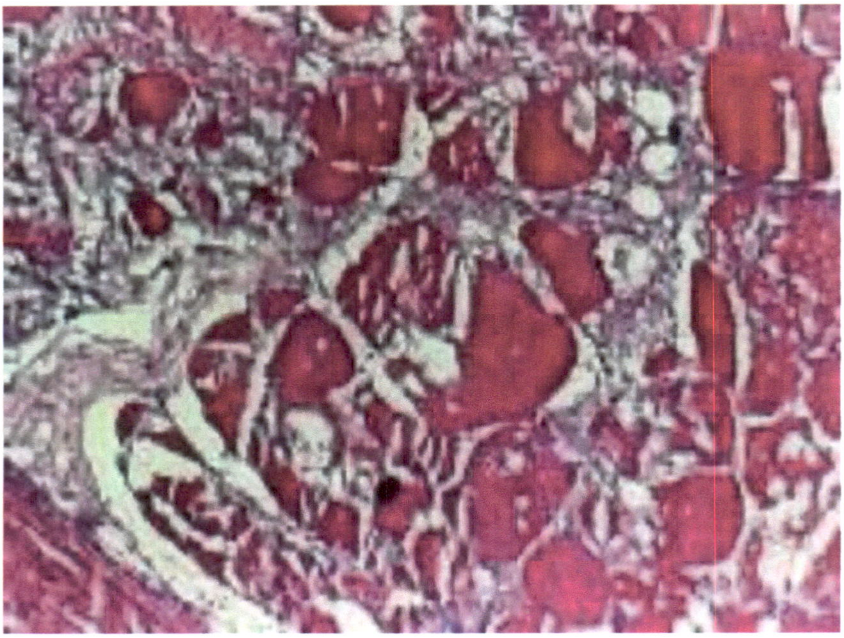

Fig. (13). Mild sarcolysis in a moderate EUS lesion.

Histopathology

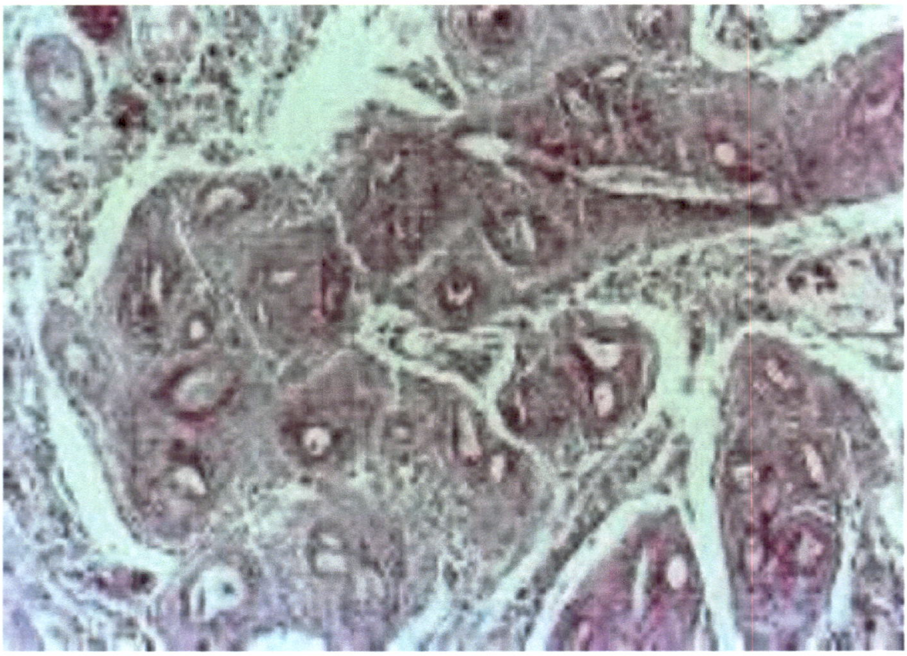

Fig. (14). Severe mycotic granulomas in a muscle lesion from an advanced case of EUS.

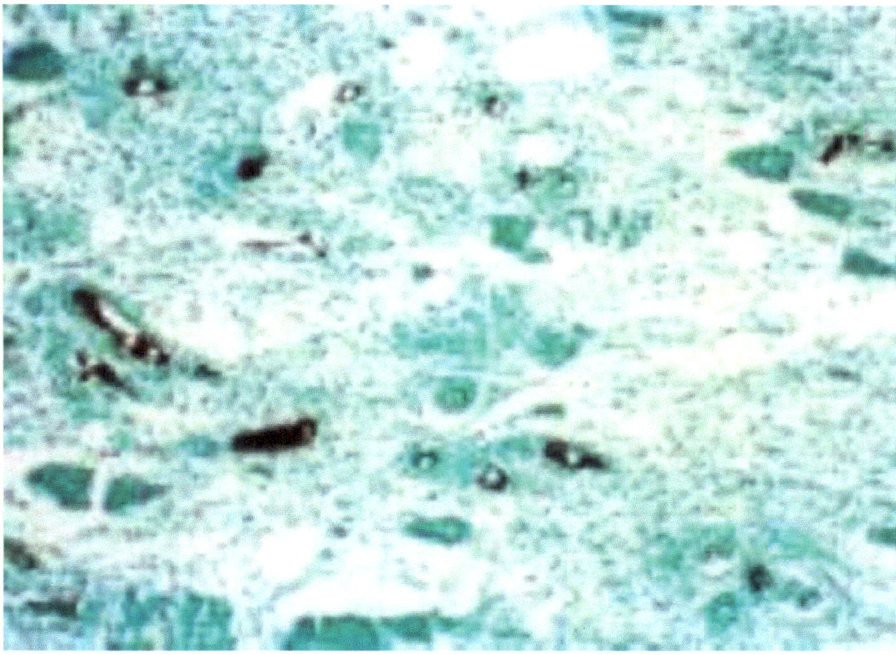

Fig. (15). Fungal hyphae (stained black) within mycotic granulomas in a muscle lesion from an advanced case of EUS. (Grocott's stain).

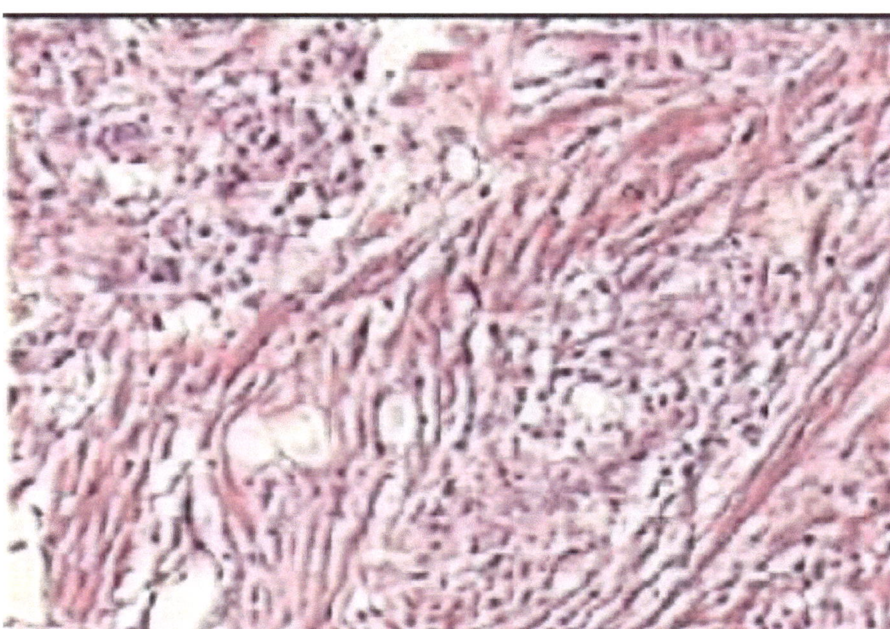

Fig. (16). Fungal hyphae in the kidney of EUS-infected fish.

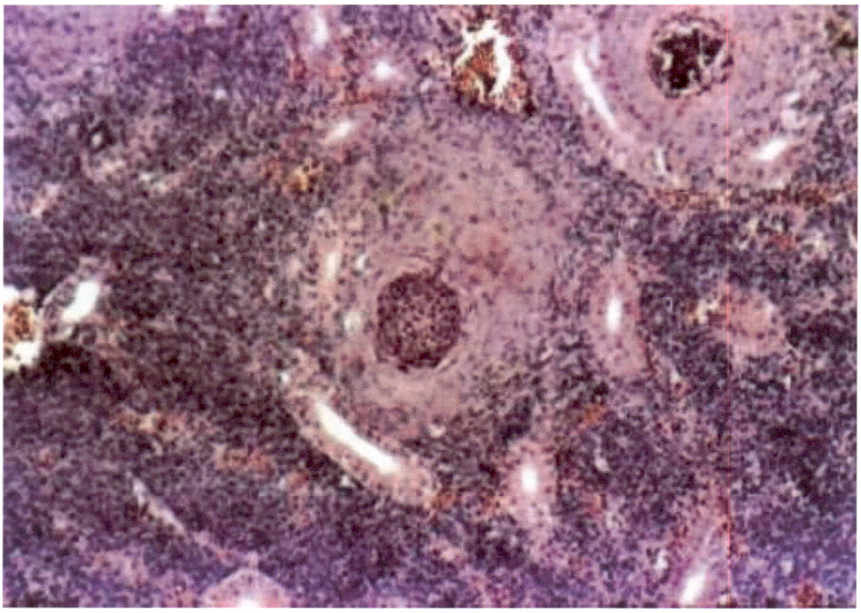

Fig. (17). Unrelated granulomas (left and right sides) and mycotic granulomas (centre) in the kidney of EUS-infected fish.

EUS in India

Fig. (18). EUS-affected Channa striatus from Dharmasagar Reservoir in Warangal, Telangana, India.

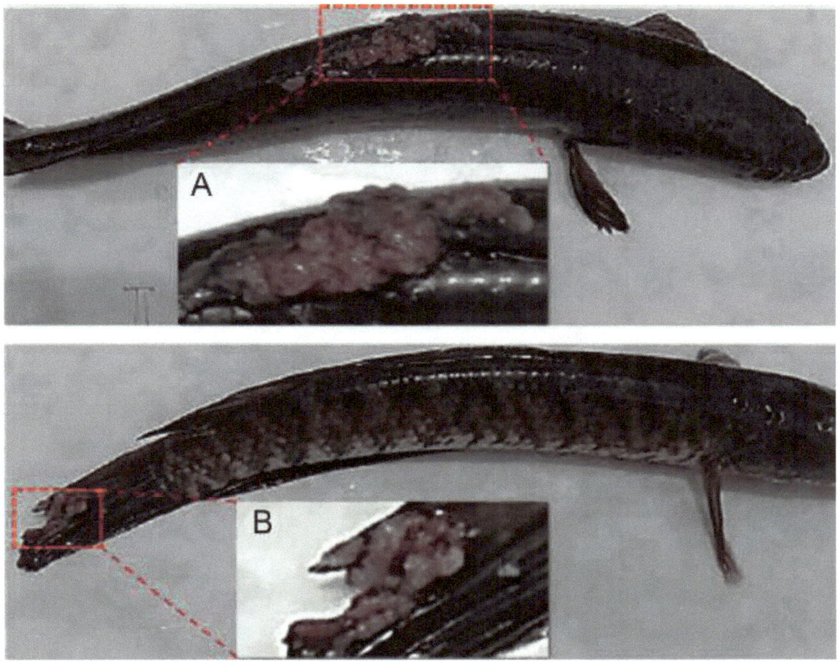

Fig. (19). EUS-affected Channa striatus from Dharmasagar Reservoir in Warangal, Telangana, India.

Fig. (20). EUS-affected Channa striatus, caught in the Bandham Lake, Warangal, Telangana, India.

Fig. (21). Dermal ulcer on a Channa striatus. The pale lesions on the ulcer surface are granulomas, which have formed in response to invasion of skeletal muscle by *A. invadans*.

Investigation Protocol

The protocol for an outbreak investigation comprises nine fundamental steps. Not all steps are invariably present in every investigation, nor do they consistently adhere to the same order. In practice, multiple steps will be executed concurrently.

THE NINE FUNDAMENTAL STEPS ARE

Development of a Diagnostic Methodology

The preliminary provisional diagnosis of an EUS epidemic is typically determined by the species of fish involved, clinical manifestations, gross pathology, and potentially seasonality. Laboratory tests should be conducted whenever feasible to confirm the provisional diagnosis. Due to the extended duration of certain laboratory processes (*e.g.*, histopathology, fungal isolation), the execution of control measures is frequently predicated on a provisional diagnosis.

Define a "Case."

Depending on the nature of the inquiry, a EUS case may involve either a single affected fish or a collective group, such as the population within an impacted pond. A pertinent case definition at the individual animal level could be 'a fish exhibiting necrotising granulomatous dermatitis and myositis linked to extremely invasive non-septate fungal hyphae'.

Verify the Occurrence of an Outbreak

This procedure may appear superfluous; yet, it is often essential, especially in regions where EUS is already endemic. The condition may be anticipated to manifest at low prevalence during specific periods; nevertheless, even a slight increase in prevalence, particularly if ulceration is extensive and/or toxigenic A is involved, is concerning. Hydrophila is prevalent and could result in significant output losses if not identified promptly. Furthermore, cutaneous ulceration induced by various treatments is prevalent in fish populations and is frequently challenging to differentiate macroscopically from EUS. Laboratory verification of an EUS diagnosis is typically required.

Characterize the outbreak in relation to temporal factors, impacted and unaffected fish, and geographical location. From an epidemiological perspective, it is crucial to delineate the outbreak concerning the aforementioned three factors. This characterization must be conducted to facilitate the development of hypotheses concerning the source, mode of transmission, and duration of the outbreak. The information is structured to seek answers to the following types of inquiries:

DURATION

What is the Specific Duration of the Outbreak?

Considering the diagnosis, what is the likely duration of exposure?

Is the outbreak predominantly a common source, propagated, or a combination of both?

Ichthyic Species

Do specific attack rates vary based on particular qualities of fish? Which groups have the highest and lowest attack rates?

Location

What are the key characteristics of the geographical distribution of cases? What are the pertinent attack rates?

Duration

Three fundamental time spans are utilized to characterize disease temporal patterns: the epidemic phase, which varies in length based on the specific epidemic; a 12-month interval to delineate seasonal patterns; and an indeterminate duration of years to define long-term trends. Understanding seasonal patterns and long-term trends is crucial for determining the existence of an epidemic in the current timeframe and for forecasting future epidemics.

The temporal dynamics of an outbreak are delineated through its epidemic curve. The epidemic curve is a graphical representation depicting the start of illness cases, presented as either a bar graph or a frequency polygon. The initial case detected in a specific outbreak is termed the 'index' case. Infectious disorders like EUS can benefit from information regarding the index case to ascertain the outbreak's source.

An epidemic curve typically comprises four, and occasionally five, parts.

- The endemic threshold
- An elevated branch
- A summit or flat expanse
- A declining branch
- A subordinate apex

The gradient of the ascending branch may signify the nature of exposure (propagating or common source) or the transmission mode and incubation duration. If transmission is swift and the incubation period is brief, the ascending branch will be steeper compared to scenarios of slow transmission or prolonged incubation periods.

The dimensions of the plateau and the gradient of the descending branch correlate with the availability of susceptible animals, which is influenced by various factors, including stocking densities, the fluctuating significance of different transmission mechanisms, and the ratio of resistant or immune fish within the at-risk population.

The selected time interval for graphing cases is crucial for interpreting the epidemic curve. The time interval must be determined based on the incubation or latency period and the duration over which the cases are disseminated. A frequent mistake in this context is choosing an excessively lengthy time span, which may mask nuanced variations in temporal patterns. A basic guideline is to establish the interval between one-eighth and one-quarter of the incubation period.

Consequently, for EUS, which has an incubation period of roughly 10 days, the incidence/prevalence in the population should be assessed every 2 days.

Outbreaks are categorized as either 'common source' (cases arising from exposure to a shared source, such as intoxications) or 'propagated source' (transmission from animal to animal, as seen in most infectious diseases). In certain EUS epidemics, both types of sources may be implicated, with initial cases arising from exposure to a common source (such as polluted water or equipment) and secondary cases stemming from fish-to-fish transmission.

The duration of an outbreak is determined by:

- The quantity of susceptible animals exposed to an infection source that subsequently get infected;
- The duration during which vulnerable animals are exposed to the infection source.
- The lowest and maximum incubation durations of the disease.

Ichthyic Species

While the term 'fish' is employed here, the expressions 'cases' and 'non-cases' should be utilized to encompass broader definitions, wherein 'cases' may refer to ponds, farms, *etc.*

Species, age, sex, and geographical origin are frequently correlated with differing illness risk. Nonetheless, it is important to recognize that fish-level patterns may be intricately associated with temporal and geographic patterns of illness.

To delineate illness trends by fish types, it is essential to specify the metrics of disease frequency employed in outbreak investigations. The primary metric for assessing disease frequency during outbreaks is the 'Attack Rate' (AR). An assault rate is a specific type of incidence rate characterized by a brief observation time. The attack rate is calculated by dividing the number of illness cases by the number of animals at risk at the onset of the epidemic. To analyze various risk factors for the disease, it is essential to compute the attack rates unique to each factor. For instance, assume there were fatalities attributed to suspected EUS in a pond, and it seemed that smaller fish were at a higher risk of contracting EUS compared to larger fish. We may perform the subsequent calculations:

Numerical Value with EUS

For little fish, AR1 = ——————————

Total number of small fish

Numerical Value with EUS

For substantial fish, AR2 = ——————————

Total Number of Huge Fish

Assuming hypothetical figures, 300 out of 1000 little fish and 100 out of 1000 large fish in the pond exhibited EUS during the outbreak. The attack rates are 30% and 10%, indicating that small fish were three times more predisposed to acquire EUS compared to large fish. This finding could support the concept that nutritionally stressed fish are more vulnerable to EUS.

The subsequent section delineates formal methodologies for comparing attack rates among groups of fish exhibiting diverse features.

Utilizing EUS, it is essential to estimate the overall fish populations across several hypothetical risk categories by representative sampling. Another issue may pertain to the calculation of mortality figures and the timing of deaths. In the investigation of an EUS outbreak, it is hypothesized that an epidermal injury, such as rhabdovirus infection, precedes the infection with A. During the outbreak, the analysis of fish samples at various intervals should aim to delineate the several lesion types and assess the rhabdovirus infection status of each fish. Therefore, the 'spectrum' of sickness is significant in this context.

Location

Characterizing the outbreak by location may facilitate the identification of its cause. For aquaculture, this may entail examining the patterns across various ponds. It is frequently advantageous to examine time and location concurrently. This can be accomplished by delineating a schematic of the ponds and documenting the dates of case occurrences. This diagram may indicate if the outbreak is of a common cause or a propagating nature. Spot maps are advantageous for extensive epidemics.

Analyzing the Data

Factor-specific attack rates for variables such as species, age, sex, pond, and management system are computed and organized in an 'attack rate table.' A theoretical example is presented below for EUS, wherein size suggests potential nutritional stress.

The attack rates in the table above are presented as percentages. The penultimate column represents the difference in assault rates, sometimes referred to as 'Attributable Risk,' while the final column denotes 'Relative Risk,' which is the ratio of the attack rates.

A greater disparity in attack rates and relative risk signifies a more significant role of the given factor in elevating illness risk. The study becomes increasingly intricate when attempting to disentangle the interactions and confounding variables among elements. Stratified and multivariate analyses are employed to examine these occurrences; however, these methods exceed the scope of this manual.

In the aforementioned hypothetical epidemic scenario, it is important to highlight that little fish had a threefold increased likelihood of having EUS compared to medium-sized fish, and a sixfold increased likelihood compared to large fish. Moreover, medium-sized fish faced twice the danger as larger fish. The dose-response relationship between size and attack rate would support the concept that nutritional stress, as indicated by size, is a 'component cause' of EUS.

Provisional Hypotheses

Working hypotheses for further inquiry are formulated based on the analysis of temporal, spatial, and ichthyological data. These may pertain to one or more of the following:

- determining if the outbreak is of common cause or proliferating nature;
- Whether a common source exists, be it a singular point or several exposures;
- The mode of transmission - touch, vehicle, or vector.

Every theory must align with all the data.

Corrective measures may be implemented based on the most plausible theories.

Rigorous Monitoring

This encompasses clinical, pathological, and microbiological assessments, along with evaluations of water quality data and recent meteorological information. Epidemiologic follow-up will encompass comprehensive analyses of this data and the investigation for further instances on alternative premises. Flowcharts depicting the management and movement of fish, water, and equipment may be necessary as part of this process.

Transmission trials may be required when additional infectious organisms, such as bacteria or ectoparasites, are recognized as contributing factors to the outbreak.

Management and Mitigation

It is hoped that the inquiry will facilitate the cessation of the outbreak and that the insights obtained will mitigate the risk of similar events in the future. Strategies to halt the pandemic must be implemented promptly, even in the absence of definitive evidence.

Significant additional research is necessary before viable strategies for addressing EUS epidemics in ponds can be established. A comprehensive examination of several outbreaks may yield significant insights into potentially critical 'component' causes.

Documentation

In the case of minor outbreaks, this may involve a concise dialogue with the farm manager, highlighting essential characteristics and necessary measures to avert additional incidents. It is advisable to consistently generate a written report to ensure a permanent record of occurrences for future reference. In the case of significant epidemics, results ought to be disseminated in scholarly publications.

For comprehensive studies, the report must include the following sections: background, methodology, results, hypotheses, financial implications (if applicable), suggestions, and appendices with laboratory reports, among other materials.

CONCLUSION

The Epizootic Ulcerative Syndrome (EUS) presents a considerable threat to global aquaculture, impacting numerous fish species. The effective management of EUS depends on a synthesis of precise diagnostic methods, preventive actions, and control strategies. Advanced diagnostic instruments, such as molecular methods like PCR, along with histopathological and histochemical analyses, have improved the early detection and surveillance of the disease. Preventive approaches, including enhancing water quality, regulating environmental stressors, and implementing biosecurity standards, are crucial in reducing epidemics. Molecular diagnostics and environmental monitoring are crucial for prompt response. Control strategies, encompassing antifungal therapies, vaccine development, and the amalgamation of conventional and contemporary management techniques, are crucial for alleviating the effects of EUS. Ongoing research and collaboration among stakeholders, including fish farmers, researchers, and policymakers, are essential for promoting sustainable aquaculture methods. Through the implementation of comprehensive and proactive methods, the aquaculture sector can effectively address EUS, safeguarding fish health, minimizing economic losses, and fostering environmental sustainability. Continued investigation into disease processes, host-pathogen interactions, and innovative therapeutic strategies will facilitate enduring solutions to EUS. Epizootic Ulcerative Syndrome (EUS) remains a considerable menace to aquaculture, resulting in major economic detriment and impacting aquatic biodiversity. Effective management of EUS necessitates a comprehensive approach that includes advanced diagnostic methods, strong prevention strategies, and dependable control measures.

AI and ML-driven preventive techniques represent a significant transformation in aquaculture disease management. Predictive modeling facilitates the early detection of outbreak hazards, whereas optimization methods guarantee sustainable and efficient agricultural operations. Decision support systems facilitate equitable access to sophisticated technology, enabling farmers and aquaculture managers to proactively address EUS hazards. These innovations collectively improve resilience, productivity, and sustainability in global aquaculture systems. The management of Epizootic Ulcerative Syndrome (EUS) in aquaculture is poised for a transformation through the implementation of Artificial Intelligence (AI) and Machine Learning (ML). These technologies provide robust solutions for the issues associated with disease detection, prevention, and control, hence maintaining the sustainability of aquaculture activities.

REVOLUTIONIZING EUS MANAGEMENT WITH AI AND ML

Artificial Intelligence and Machine Learning facilitate a transformative change in Endoscopic Ultrasound management by implementing:

Improved Diagnostic Precision:

AI-driven picture recognition and predictive models provide swift and dependable disease identification, exceeding conventional techniques in both speed and accuracy. Data-Driven Prevention, predictive analytics, and decision-support systems provide proactive interventions utilizing environmental and historical data, hence mitigating outbreak risks. Efficient Control Strategies AI-driven treatment recommendation systems, together with continuous monitoring, enhance interventions while reducing resource consumption and environmental impact. These inventions enhance fish health and farm productivity while also addressing global issues such as food security and sustainable aquaculture. The significance of Technology Adoption: To harness the potential of AI and ML, it is essential to overcome adoption barriers and cultivate a collaborative ecosystem. Empowering small-scale farmers through accessible, cost-effective technologies tailored to their needs can democratize the advantages of AI, facilitating widespread adoption. Advancing Sustainability: AI technologies enhance resource efficiency, diminish waste, and lessen dependence on detrimental methods, harmonizing aquaculture with environmental conservation objectives, enhancing resilience by incorporating real-time monitoring, early warning systems, and sophisticated analytics. Aquaculture operations may more effectively respond to problems such as climate change and disease outbreaks. The integration of AI and ML is essential for the future of sustainable aquaculture. By embracing these technologies, governments, researchers, technology developers, and farmers can advance the aquaculture industry towards a future where EUS and other illnesses are effectively managed, safeguarding the health of aquatic ecosystems and the lives of millions reliant on them. Artificial Intelligence and Machine Learning enable harmonizing productivity with sustainability, facilitating a robust and thriving aquaculture industry.

ACKNOWLEDGEMENT

The authors would like to thank Dr. P. Srinivas, Plant Pathology and Microbiology Laboratory, Department of Biotechnology, Kakatiya University, Warangal, and Dr. M. Esthari, Department of Zoology, Kakatiya University, Warangal, for their continuous support and inspiration, and for providing the necessary facilities for the work. This publication was prepared in collaboration between the Department of Zoology, Microbiology, and Biotechnology of Kakatiya University, Warangal, Telangana, India.

REFERENCES

References

AAHRI, ACIAR, IoA and NACA (1997) Epizootic ulcerative syndrome (EUS) of fishes in Pakistan. A report of the findings of an ACIAR/ DFID-funded mission to Pakistan. 9-19 March 1997 *AAHRI.*

ADB/NACA (1991) Fish Health Management in Asia-Pacific. Report on a Regional Study and Workshop on Fish Disease and Fish Health Management *ADB Agriculture Department Report Series No 1 Network of Aquaculture Centres in Asia-Pacific, Bangkok.*

Ahmed, M & Rab, MA (1995) Factors affecting outbreaks of epizootic ulcerative syndrome in farmed fish in Bangladesh. *J Fish Dis,* 18, 263-72.
[http://dx.doi.org/10.1111/j.1365-2761.1995.tb00301.x]

Ahmed, N & Allison, EH (2021) Use of predictive analytics for disease outbreak control in aquaculture. *Aquacult Res,* 52, 2158-71.

Ahmed, GU & Hoque, A (1998) Mycotic involvement in epizootic ulcerative syndrome of freshwater fishes of Bangladesh: a histopathological study. *Proceedings of the Fifth Asian Fisheries Forum.* Chiang Mai, ThailandAsian Fisheries Society, Manila.

Ahne, W, Jørgensen, PEV, Olesen, NJ & Wattanavijarn, W (1988) Serological examination of a rhabdovirus isolated from snakehead (*Ophicephalus striatus*) in Thailand with ulcerative syndrome. *J Appl Ichthyology,* 4, 194-6.
[http://dx.doi.org/10.1111/j.1439-0426.1988.tb00562.x]

Alderman, DJ & Polglase, JL (1988) Pathogens, parasites and commensals. In: Holdich, D.M., Lowery, R.S., (Eds.), *Freshwater Crayfish: Biology* Management and Exploitation 167-212.

Anon. (1992) Notes on mycological procedure. Workshop on Mycological Aspects of Fish and Shellfish Disease, Bangkok, Thailand, January 1992. Aquatic Animal Health Research Institute.

Avtalion, RR, Wichkovsky, A & Katz, D (1980) Regulatory effect of temprature on specific supression and enhancement of the humoral response in fish. In: Manning, M.J., (Ed.), *Phylogeny of Immunological Memory* Elsevier, Amsterdam 113-21.

Balasuriya, LKSW (1994) Epizootic ulcerative syndrome in fish in Sri Lanka, country status report. In: Roberts, R.J., Campbell, B., MacRae, I.H., (Eds.), *Proceedings of the ODA Regional Seminar on Epizootic Ulcerative Syndrome* 39-47.Aquatic Animal Health Research Institute, Bangkok.

Barua, G. (1990) Bangladesh report. In: Phillips, M.J. and Keddie, H.G. (Eds) Regional Research Programme on Relationships Between Epizootic Ulcerative Syndrome in Fish and the Environment. A Report on the Second Technical Workshop, 13-26 August 1990. Network of Aquaculture Centres in Asia-Pacific, Bangkok.

Barredo Arrieta, A (2020) Explainable artificial intelligence in aquaculture: Advances and applications. *Inf Fusion,* 58, 82-96.
[http://dx.doi.org/10.1016/j.inffus.2019.12.012]

Barua, G (1994) The status of epizootic ulcerative syndrome of fish of Bangladesh. In: Roberts, R.J., Campbell, B., MacRae, I.H., (Eds.), *Proceedings of the ODA Regional Seminar on Epizootic Ulcerative Syndrome* 13-20.Aquatic Animal Health Research Institute, Bangkok.

Bhaumik, U, Pandit, PK & Chatterjee, JG (1991) Impact of epizootic ulcerative syndrome on the fish yield, consumption and trade in West Bengal. *J Inland Fish Soc India,* 23, 45-51.

Bly, JE & William Clem, L (1991) Temperature-mediated processes in teleost immunity: In vitro immunosuppression induced by in vivo low temperature in channel catfish. *Vet Immunol Immunopathol,* 28, 365-77.
[http://dx.doi.org/10.1016/0165-2427(91)90127-X] [PMID: 1949572]

Bondad-Reantaso, M.G., Lumanlan, S.C., Natividad, J.M. and Phillips, M.J. (1992) Environmental monitoring of the epizootic ulcerative syndrome (EUS) in fish from Munoz, Nueva Ecija in the Philippines. In: Shariff, M., Subasinghe, R.P. and Arthur, J.R. (Eds) Diseases in Asian Aquaculture 1. pp. 475-490. Fish Health Section, Asian Fisheries Society, Manila.

Bondad-Reantaso, M.G., Paclibare, J.O., Lumanlan-Mayo, S.C. and Catap, E.S. (1994) EUS outbreak in the Philippines: a country report. In: Roberts, R.J., Campbell, B. and MacRae, I.H. (Eds), Proceedings of the ODA Regional Seminar on Epizootic Ulcerative Syndrome, 25-27 January 1994. pp. 61-67. Aquatic Animal Health Research Institute, Bangkok.

Boonyaratpalin, S. (1989a) A report on epizootic ulcerative sysndrome of fish in India. Paper code NACA/WP/89/80, Network of Aquaculture Centres in Asia, Bangkok.

Bui, DT (2020) Machine learning techniques for predicting fish diseases in aquaculture systems. *Aquacult Res,* 51, 1616-30.

Burkholder, JM, Noga, EJ, Hobbs, CH, Glasgow, HB, Jr & Smith, SA (1992) New ‘phantom’ dinoflagellate is the causative agent of major estuarine fish kills. *Nature,* 358, 407-10.
[http://dx.doi.org/10.1038/358407a0] [PMID: 1641022]

Callinan, R.B. (1994b) Red spot disease - EUS in Australia. In: Roberts, R.J., Campbell, B. and MacRae, I.H. (Eds) Proceedings of the ODA Regional Seminar on Epizootic Ulcerative Syndrome, 25-27 January 1994. pp. 82-88. Aquatic Animal Health Research Institute, Bangkok.

Callinan, R.B. (1997) Pathogenesis of red spot disease (epizootic ulcerative syndrome) in estuarine fish in eastern Australia and the Philippines. Ph.D. Thesis, University of Queensland, pp. 232.

Chakraborty, C (2021) AI-driven prophylactic measures for sustainable aquaculture practices. *AI in Fisheries Management,* 2, 14-25.

Cheng, X (2023) AI-enabled biofilm control strategies to mitigate EUS outbreaks. *Appl Microbiol Biotechnol,* 107, 563-78.

Callinan, RB & Keep, JA (1989) Bacteriology and parasitology of red spot disease in sea mullet, *Mugil cephalus* L., from eastern Australia. *J Fish Dis,* 12, 349-56.
[http://dx.doi.org/10.1111/j.1365-2761.1989.tb00323.x]

Callinan, R.B. and Keep, J.A. (1989) Bacteriology and parasitology of red spot disease in sea mullet, Mugil cephalus L., from eastern Australia. Journal of Fish Diseases 12, 349-356.

Callinan, R.B., Paclibare, J.O., Reantaso, M.B., Lumanlan-Mayo, S.C., Fraser, G.C. and Sammut, J. (1995b) In: Shariff, M., Arthur, J.R. and Subasinghe, R.P. (Eds) EUS outbreaks in estuarine fish in Australia and the Philippines: associations with acid sulphate soils, rainfall and Aphanomyces. Diseases in Asian Aquaculture II. pp. 291-298. Fish Health Section, Asian Fisheries Society, Manila.

Cavalier-Smith, T (1987) The origin of Fungi and pseudofungi. In: Rayner, A.D.M., Brasier, C.M., Moore, D., (Eds.), *Evolutionary Biology of the Fungi* Cambridge University Press 339-53.

Celio, DA & Padgett, DE (1989) An improved method of quantifying water mould spores in natural water columns. *Mycologia,* 81, 459-60.
[http://dx.doi.org/10.1080/00275514.1989.12025768]

Chakrabarty, AN & Dastidar, SG (1991) Repeated isolation of chemoautotrophic nocardioform bacteria from fish epizootic ulcerative syndrome. *Indian J Exp Biol,* 29, 623-7.
[PMID: 1794852]

Chinabut, S (1994) EUS in Thailand. In: Roberts, R.J., Campbell, B., Macrae, I.H., (Eds.), *Proceedings of the ODA Regional Seminar on Epizootic Ulcerative Syndrome* 25-27 January 199458-60.Aquatic Animal Health Research Institute, Bangkok, Thailand.

Chinabut, S, Roberts, RJ, Willoughby, GR & Pearson, MD (1995) Histopathology of snakehead, *Channa striatus* (Bloch), experimentally infected with the specific *Aphanomyces* fungus associated with epizootic ulcerative syndrome (EUS) at different temperatures. *J Fish Dis,* 18, 41-7.
[http://dx.doi.org/10.1111/j.1365-2761.1995.tb01264.x]

Chondar, S.L. and Rao, P.S. (1996) Epizootic ulcerative syndrome disease to fish and its control: a review. World Aquaculture 1996, Queen Sirikit National Convention Centre, Bangkok, 29 Jan - 2 Feb 1996. Book of Abstracts. P. 77. World Aquaculture Society.

Chulalongkorn University. (1983) The Symposium on Fresh Water Fishes Epidemic: 1982-1983, Chulalongkorn University, Bangkok, 23-24 June 1983.

Chulalongkorn University. (1985) Proceedings of the Technical Conference on Living Aquatic Resources, Chulalongkorn University, Bangkok, 7- 8 March 1985.

Chulalongkorn University. (1987) Proceedings of the Second Seminar on Living Aquatic Resources, Chulalongkorn University, Bangkok, 17- 18 December 1987.

Clark, S (1991) Snakeheads. The species and their distribution. *Aquar Pondkeep,* 56, 17-22.

Coates, D, Nunn, MJ & Uwate, KR (1989) Epizootic Ulcerative Disease of freshwater fish in Papua New Guinea. *Sci New Guinea,* 15, 1-11.

Costa, HH & Wijeyaratne, MJS (1989) Epidemiology of the epizootic ulcerative syndrome occurring for the first time among fish in Sri Lanka. *J Appl Ichthyology,* 5, 48-52.
[http://dx.doi.org/10.1111/j.1439-0426.1989.tb00570.x]

Cruz-Lacierda, E.R. and Shariff, M. (1995) Experimental transmission of epizootic ulcerative syndrome (EUS) in snakehead (*Ophicephalus striatus*). In: Shariff, M., Arthur, J.R. and Subasinghe, R.P. (Eds) Diseases in Asian Aquaculture II. pp. 327-336. Fish Health Section, Asian Fisheries Society, Manila.

Das, MK (1994) Outbreak of the fish disease, epizootic ulcerative syndrome in India - an overview. In: Roberts, R.J., Campbell, B., Macrae, I.H., (Eds.), *Proceedings of the ODA Regional Seminar on Epizootic Ulcerative Syndrome* 25-27 January 199421-38.Aquatic Animal Health Research Institute, Bangkok, Thailand.

Das, MK & Das, RK (1993) A review of the fish disease epizootic ulcerative syndrome in India. *Environ Ecol,* 11, 134-45.

Dykstra, MJ, Noga, EJ, Levine, JF, Moye, DW & Hawkins, JH (1986) Characterization of the Aphanomyces species involved with ulcerative mycosis (UM) in menhaden. *Mycologia,* 78, 664-72.

[http://dx.doi.org/10.1080/00275514.1986.12025303]

Egusa, S & Masuda, N (1971) A new fungal of Plecoglossus altivelis. *Fish Pathol,* 6, 41-6.
[http://dx.doi.org/10.3147/jsfp.6.41]

FAO (1986) *Report of the expert consultation on ulcerative fish diseases in the Asia-Pacific region (TCP/RAS/4508) Bangkok 5-9 August 1986.* FAO, Regional Office for Asia and the Pacific, Bangkok.

Fowles, B (1976) Factors affecting growth and reproduction in selected species of Aphanomyces. *Mycologia,* 68, 1221-32.
[http://dx.doi.org/10.1080/00275514.1976.12020011] [PMID: 13300]

Fraser, GC, Callinan, RB & Calder, LM (1992) *Aphanomyces* species associated with red spot disease: an ulcerative disease of estuarine fish from eastern Australia. *J Fish Dis,* 15, 173-81.
[http://dx.doi.org/10.1111/j.1365-2761.1992.tb00651.x]

Frerichs, GN (1995) Viruses associated with the epizootic ulcerative syndrome (EUS) of fish in South-East Asia. *Vet Res,* 26, 449-54.
[PMID: 8581021]

Frerichs, GN, Hill, BJ & Way, K (1989) Ulcerative disease rhabdovirus: cell-line susceptibility and serological comparison with other fish rhabdoviruses. *J Fish Dis,* 12, 51-6.
[http://dx.doi.org/10.1111/j.1365-2761.1989.tb01290.x]

Frerichs, GN, Millar, SD & Alexander, M (1989) Rhabdovirus infection of ulcerated fish in South-East Asia. In: Ahne, W., Kurstak, E., (Eds.), *Viruses of Lower Vertebrates* Springer-Verlag, Berlin 396-410.
[http://dx.doi.org/10.1007/978-3-642-83727-2_36]

Frerichs, GN, Millar, SD & Chinabut, S (1993) Clinical response of snakeheads (*Ophicephalus striatus*) to experimental infection with snakehead fish rhabdovirus and snakehead cell line retrovirus. *Aquaculture,* 116, 297-301.
[http://dx.doi.org/10.1016/0044-8486(93)90414-T]

Guizhen, J. (1990) Peoples Republic of China report. In: Phillips, M. J. and Keddie, H. G. (Eds) Regional Research Programme on Relationships Between Epizootic Ulcerative Syndrome in Fish and the Environment. A Report on the Second Technical Workshop, 13-26 August 1990. Network of Aquaculture Centres in Asia-Pacific, Bangkok.

Haines, AK (1983) Fish fauna and ecology. In: Petr, T., (Ed.), *The Purari tropical environment of high rainfall river basin* Dr W. Junk 367-84.
[http://dx.doi.org/10.1007/978-94-009-7263-6_20]

Hassanien, AE & Mahdy, YB (2019) Fuzzy logic systems for diagnosing epizootic ulcerative syndrome in aquaculture. *Comput Electron Agric,* 162, 276-85.

Hanjavanit, C, Suda, H & Hatai, K (1997) Mycotic granulomatosis found in two species of ornamental fishes imported from Singapore. *Mycoscience,* 38, 433-6.
[http://dx.doi.org/10.1007/BF02461684]

Hatai, K (1980) *Studies on pathogenic agents of saprolegniasis in fresh water fishes Special Report of Nagasaki Prefectural Institute of Fisheries No8.* Matsugae-cho, Nagasaki, Japan.

Hatai, K (1994) Mycotic granulomatosis in ayu (Plecoglossus altivelis) due to Aphanomyces piscicida. In: Roberts, R.J., Campbell, B., MacRae, I.H., (Eds.), *Proceedings of the ODA Regional Seminar on Epizootic Ulcerative Syndrome.* 25-27 January 1994Aquatic Animal Health Research Institute, Bangkok..

Hatai, K & Egusa, S (1978) Studies on the pathogenic fungus of mycotic granulomatosis - II. Some of the note on the MG-fungus. *Fish Pathol,* 13, 85-9.
[http://dx.doi.org/10.3147/jsfp.13.85]

Hatai, K, Egusa, S, Takahashi, S & Ooe, K (1977) Study on the pathogenic fungus of mycotic granulomatosis - I. Isolation and pathogenicity of the fungus from cultured ayu infected with the disease. *Fish Pathol,* 12, 129-33.
[http://dx.doi.org/10.3147/jsfp.12.129]

Hatai, K, Nakamura, K, Rha, SA, Yuasa, K & Wada, S (1994) Aphanomyces infection in dwarf gourami (Colisa lalia). *Fish Pathology*, 29(2), 95-99.

Hedrick, R.P., Eaton, W.D., Fryer, J.L., Groberg, W.G. Jr. and Boonyaratpalin, S. (1986) Characteristics of a birnavirus isolated from cultured sand goby, Oxyeleotris marmoratus. *Dis Aquat Organ*, 1, 219-25.

Hoshina, T, Sano, T & Sunayama, M (1960) Studies on the saprolegniasis of eel. *J Tokyo Univ Fish,* 47, 59-79.

Humphrey, J., Arthur, J.R., Subasinghe, R.P. and Phillips, M.J. (1997) Aquatic Animal Quarantine and Health Certification in Asia. Proceedings of the Regional Workshop on Health and Quarantine Guidelines for the Responsible Movement (Introduction and Transfer) of Aquatic Organisms. Bangkok. 28 January 1996. (FAO Fisheries Technical Paper No. 373). FAO, Rome.

ICSF (1992) *Enigma of EUS Consultation on Epizootic Ulcerative Syndrome Vis-A-Vis the Environment and the People 25-26 May 1992.* Trivandrum, Kerala. International Collective in Support of Fishworkers, Madras, India.

Jhingran, AG & Das, MK (1990) *Epizootic ulcerative syndrome in fishes Bulletin of the Central Inland Capture Fisheries Research Institute (No65).* CIFRI, Barrackpore, India.

Jothy, A.A. (1981) Preliminary report on the outbreak of wabak kudison freshwater fish in paddy growing areas in Peninsular Malaysia. Report to the Ministry of Agriculture, Malaysia. December 1981. 17.

Kanchanakhan, S (1996) Epizootic ulcerative syndrome (EUS): a new look at the old story. *AAHRI Newsletter*, 5, 2-3.

Kanchanakhan, S. (1996b) Field and laboratory studies on rhabdoviruses associated with epizootic ulcerative syndrome (EUS) of fishes. Ph.D. Thesis, University of Stirling, Scotland. 278.

Kumar, S & Mohanty, A (2020) Machine learning in risk assessment of EUS outbreaks in freshwater aquaculture. *Aquacult Rep,* 18, 100499.

Kasornchandra, J, Engelking, HM, Lannan, CN, Rohovec, JS & Fryer, JL (1992) Characteristics of three rhabdoviruses from snake-head fish *Ophicephalus striatus. Dis Aquat Organ,* 13, 89-94.
[http://dx.doi.org/10.3354/dao013089]

Karunasagar, I., Sugumar, G., and Karunasagar, I. (1995) Virulence characters of Aeromonas spp isolated from EUS-affected fish. In: Shariff, M., Arthur, J.R. and Subasinghe, R.P. (Eds), Diseases in Asian Aquaculture II pp. 307-314. Fish Health Section, Asian Fisheries Society, Manila.

Khan, MH, Marshall, L, Thompson, KD, Campbell, RE & Lilley, JH (1998) Susceptibility of five fish species (Nile tilapia, rosy barb, rainbow trout, stickleback and roach) to intramuscular injection with the Oomycete fish pathogen, Aphanomyces invadans. *Bull Eur Assoc Fish Pathol,* 18, 192-7.

Khulbe, RD, Joshi, C & Bisht, GS (1995) Fungal diseases of fish in Nanak Sagar, Naini Tal, India. *Mycopathologia,* 130, 71-4.
[http://dx.doi.org/10.1007/BF01103452] [PMID: 7566059]

Kumar, D, Dey, RK & Sinha, A (1991) Outbreak of epizootic ulcerative syndrome of fish in India. In: Sinha, V.R.P., Srivastava, H.C., (Eds.), *Aquaculture Productivity* Oxford and IBH Publishing Company, New Delhi 345-56.

Levine, JF, Hawkins, JH, Dykstra, MJ, Noga, EJ, Moye, DW & Cone, RS (1990) Species distribution of ulcerative lesions on finfish in the Tar-Pamlico River Estuary, North Carolina. *Dis Aquat Organ,* 8, 1-5.
[http://dx.doi.org/10.3354/dao008001]

Li, J (2019) Development of an AI-based diagnostic system for fish diseases using convolutional neural networks. *Aquacult Int,* 27, 1677-92.

Levine, JF, Hawkins, JH, Dykstra, MJ, Noga, EJ, Moye, DW & Cone, RS (1990) Epidemiology of ulcerative mycosis in Atlantic menhaden in the Tar-Pamlico River Estuary, North Carolina. *J Aquat Anim Health,* 2, 162-71.
[http://dx.doi.org/10.1577/1548-8667(1990)002<0162:EOUMIA>2.3.CO;2]

Lian, C.X. (1990) Peoples Republic of China report. In: Phillips, M.J. and Keddie, H.G. (Eds) Regional Research Programme on Relationships Between Epizootic Ulcerative Syndrome in Fish and the Environment. A Report on the Second Technical Workshop, 13-26 August 1990. Network of Aquaculture Centres in Asia-Pacific, Bangkok.

Lilley, JH & Frerichs, GN (1994) Comparison of rhabdoviruses associated with epizootic ulcerative syndrome (EUS) with respect to their structural proteins, cytopathology and serology. *J Fish Dis,* 17, 513-22.
[http://dx.doi.org/10.1111/j.1365-2761.1994.tb00246.x]

Lilley, JH & Inglis, V (1997) Comparative effects of various antibiotics, fungicides and disinfectants on Aphanomyces invaderis and other saprolegniaceous fungi. *Aquacult Res,* 28, 461-9.
[http://dx.doi.org/10.1111/j.1365-2109.1997.tb01064.x]

Lilley, JH & Roberts, RJ (1997) Pathogenicity and culture studies comparing the *Aphanomyces* involved in epizootic ulcerative syndrome (EUS) with other similar fungi. *J Fish Dis,* 20, 135-44.
[http://dx.doi.org/10.1046/j.1365-2761.1997.d01-116.x]

Lilley, JH, Hart, D, Richards, RH, Roberts, RJ, Cerenius, L & Söderhäll, K (1997) Pan-Asian spread of single fungal clone results in large scale fish kills. *Vet Rec,* 140, 653-4.
[http://dx.doi.org/10.1136/vr.140.25.653] [PMID: 9226850]

Llobrera, AT & Gacutan, RQ (1987) *Aeromonas hydrophila* associated with ulcerative disease epizootic in Laguna de Bay, Philippines. *Aquaculture,* 67, 273-8.
[http://dx.doi.org/10.1016/0044-8486(87)90211-0]

Llobrera, A. and Gacutan, R.Q. (1987) Aeromonas hydrophila associated with ulcerative disease epizootic in Laguna de Bay, Philippines. Aquaculture 67, 273-278.

Mao, R (2021) Predictive models for environmental factors linked to EUS outbreaks using AI. *Sci Total Environ,* 764, 142841.

Macintosh, DJ (1986) Environmental background to the ulcerative disease conditionin South-east Asia. In: Roberts, R.J., Macintosh, D.J., Tonguthai, K., Boonyaratpalin, S., Tayaputch, N., Phillips, M.J., Millar, S.D., (Eds.), *Field and Laboratory Investigations into Ulcerative Fish Diseases in the Asia-Pacific Region*

Technical Report of FAO Project TCP/RAS/4508 61-113.

Marimuthu, K (2021) AI for disease resilience and optimization in aquaculture systems. *Rev Aquacult,* 13, 726-38.

Matthews, M & Reynolds, JD (1990) Laboratory investigations of the pathogenicity of Aphanomyces astaci for Irish freshwater crayfish. *Hydrobiologia,* 203, 121-6.
[http://dx.doi.org/10.1007/BF00005680]

Miyazaki, T (1994) Comparison among mycotic granulomatosis, saprolegniasis and anaaki-byo in fishes: a Japanese experience. In: Roberts, R.J., Campbell, B., MacRae, I.H., (Eds.), *Proceedings of the ODA Regional Seminar on Epizootic Ulcerative Syndrome* Aquatic Animal Health Research Institute, Bangkok 253-70.

Mohan, C.V. and Shankar, K.M. (1995) Role of fungus in epizootic ulcerative syndrome of fresh- and brackishwater fishes of India: a histopathological assessment. In: Shariff, M., Arthur, J.R. and Subasinghe, R.P. (Eds). Diseases in Asian Aquaculture II. pp. 299-305. Fish Health Section, Asian Fisheries Society, Manila.

Munday, BL (1985) Ulcer disease in cod (Pseudophycis bachus) from the Tamar River. *Tasman Fish Res,* 27, 15-8.

Myers, GS (1931) On the identity of Ophicephalus and Channa, two genera of labyrinth fishes. *Peking Natural History Bulletin,* 6, 33-7.

National Workshop on Ulcerative Disease Syndrome in Fish. (1990) Proceedings of the National Workshop on Ulcerative Disease Syndrome in Fish, Calcutta, India, 6-7 March 1990.

Noga, EJ (1993) *Fungal diseases of marine and estuarine fishes Pathobiology of Marine and Estuarine Organisms* CRC Press, Inc. 85-110.

Nguyen, TT (2022) Genomic insights and machine learning in managing fungal pathogens causing EUS. *Fungal Ecol,* 55, 101125.

Noga, EJ (1994) Epidemic ulcerative diseases recently affecting estuarine fishes of the western Atlantic ocean. In: Roberts, R.J., Campbell, B., MacRae, I.H., (Eds.), *Proceedings of the ODA Regional Seminar on Epizootic Ulcerative Syndrome* 25-27 January 199489-100.Aquatic Animal Health Research Institute, Bangkok.

Noga, EJ & Dykstra, MJ (1986) Oomycete fungi associated with ulcerative mycosis in menhaden, *Brevoortia tyrannus* (Latrobe). *J Fish Dis,* 9, 47-53.
[http://dx.doi.org/10.1111/j.1365-2761.1986.tb00978.x]

Noga, EJ, Khoo, L, Stevens, JB, Fan, Z & Burkholder, JM (1996) Novel toxic dinoflagellate causes epidemic disease in estuarine fish. *Mar Pollut Bull,* 32, 219-24.
[http://dx.doi.org/10.1016/0025-326X(95)00114-3]

Noga, EJ, Wright, JF, Levine, JF, Dykstra, MJ & Hawkins, JH (1991) Dermatological diseases affecting fishes of the Tar-Pamlico Estuary, North Carolina. *Dis Aquat Organ,* 10, 87-92.
[http://dx.doi.org/10.3354/dao010087]

Ogbonna, CIC & Alabi, RO (1991) Studies on species of fungi associated with mycotic infections of fish in a Nigerian freshwater fish pond. *Hydrobiologia,* 220, 131-5.
[http://dx.doi.org/10.1007/BF00006545]

Patel, A (2023) AI-integrated management frameworks for sustainable aquaculture. *Aquaculture,* 563,

738291.

Paclibare, JO, Catap, ES & Callinan, RB (1994) Fungal isolation from EUS-affected fish in the Philippines. In: Roberts, R.J., Campbell, B., MacRae, I.H., (Eds.), *Proceedings of the ODA Regional Seminar on Epizootic Ulcerative Syndrome* Aquatic Animal Health Research Institute, Bangkok 238-43.

Palisoc, F & Aralar, E (1995) Relationships between the environment in Laguna Lake and Lake Naujan and the epizootic ulcerative syndrome (EUS) in fish. *Abstracts from the Fourth Asian Fisheries Forum* 16- 20 October 199536.Asian Fisheries Society, Manila and China.

Pearce, M. (1990) Epizootic ulcerative syndrome technical report December 1987 - September 1989. Northern Territory Department of Primary Industry and Fisheries. Fisheries Report No.22. Northern Territory, Australia. p. 82.

Phillips, M.J. (1989) A report on the NACA workshop on the regional research programme on ulcerative syndrome in fish and the environment. 20-24 March 1989. Network of Aquaculture Centres in Asia-Pacific, Bangkok.

Phillips, M.J. and Keddie, H.G. (1990) Regional Research Programme on Relationships Between Epizootic Ulcerative Syndrome in Fish and the Environment. A Report on the Second Technical Workshop, 13-26 August 1990. 133pp. Network of Aquaculture Centres in Asia-Pacific, Bangkok.

Pichyangkura, S & Tangtrongpiros, J (1985) The relationship between microscopic exam of Achlya sp. infection and characteristic of lesions *Ophicephalus striatus*. 19-23.*Proceedings of the Living Aquatic Resources* 7-8 March 1985Chulalongkorn University, Bangkok.

Pichyangkura, S & Bodhalamik, V (1983) The study of Achlya sp of fish disease in *Ophicephalus striatus*. *The Symposium on Fresh Water Fishes Epidemic* 1982-1983197-205.Chulalongkorn University, Bangkok.

Plumb, JA, Grizzle, JM & Defigueiredo, J (1976) Necrosis and bacterial infection in channel catfish (ictalurus punctatus) following hypoxia. *J Wildl Dis,* 12, 247-53.
[http://dx.doi.org/10.7589/0090-3558-12.2.247] [PMID: 933317]

Poonsuk, K, Saitanu, K, Navephap, O & Wongsawang, S (1983) Antimicrobial susceptibility testing of Aeromonas hydrophila: strain isolated from fresh water fish infection. 436-8.*The Symposium on Fresh Water Fishes Epidemic* 1982-1983Chulalongkorn University, Bangkok.

Qureshi, T.A., Chouhan, R., Prasad, Y. and Mastan, S.A. (1995) Mycological studies on EUS affected catfish, Mystus cavasius. P.38. Abstracts of the Fourth Asian Fisheries Forum. 16-20 October 1995. Asian Fisheries Society and China Society Fisheries. Beijing.

Raman, RP (1992) *EUS strikes in the brackishwaters of Chilka Lagoon in India Fish Health Section Newsletter 3(2), 3-4.* Asian Fisheries Society, Manila.

Ranjan, R & Gupta, R (2022) Role of ML-based climate prediction models in controlling aquatic diseases. *J Fish Dis,* 45, 851-60.

Rico, A (2022) Digital transformations in aquaculture disease control: Challenges and solutions. *J World Aquacult Soc,* 53, 919-38.

Rahman, T, Choudhury, B & Barman, N (1988) Role of UDS affected fish in the health of ducks in Assam: an experimental study. *Proceedings of the Symposium on Recent Outbreak of Fish Diseases in North Eastern India,* 30 December 198810.Guwahati, Assam, India.

Rattanaphani, R, Wattanavijarn, W, Tesprateep, T, Wattanodorn, S, Sukolopong, V, Vetchagarun, S &

Eongpakornkeaw, A (1983) Preliminary report on fish virus in snakehead fish. *Wetchasan Sattawaphaet,* 13, 44-50.

Reantaso, MB (1991) *EUS in brackishwaters of the Philippines Fish Health Section Newsletter 2(1), 8-9.* Asian Fisheries Society, Manila.

Reungprach, H, Boonyaratpalin, S, Supamatya, K, Kesornchandra, J, Polsheivin, W & Sadvakdee, J (1983) *Special Report of the Fish Disease Outbreak Committee in Thailand.* Ministry of Agriculture and Cooperatives, Thailand.

Riji John, K. (1997) Characterisation of reovirus-like agents associated with snakehead fish and cell culture. Ph.D. Thesis, University of Stirling, Scotland.

Roberts, RJ, Campbell, B & MacRae, IH (1994) *Proceedings of the Regional Seminar on Epizootic Ulcerative Syndrome.* 25-27 January 1994The Aquatic Animal Health Research Institute, Bangkok..

Roberts RJ, Frerichs GN, Tonguthai K, Chinabut S. Epizootic ulcerative syndrome of farmed and wild fishes. In: Muir JF, Roberts RJ, editors. Recent advances in aquaculture V. Oxford: Blackwell Science; 1994. p. 207-239.

Roberts, R.J., Macintosh, D.J., Tonguthai, K., Boonyaratpalin, S., Tayaputch, N., Phillips, M.J. and Millar, S.D. (1986) Field and laboratory investigations into ulcerative fish diseases in the Asia-Pacific region. Technical Report of FAO Project TCP/RAS/4508. Bangkok, p. 214.

Roberts, RJ, Willoughby, LG & Chinabut, S (1993) Mycotic aspects of epizootic ulcerative syndrome (EUS) of Asian fishes. *J Fish Dis,* 16, 169-83.
[http://dx.doi.org/10.1111/j.1365-2761.1993.tb01248.x]

Roberts, R.J., Willoughby, L.G. and Chinabut, S. (1993) Mycotic aspects of epizootic ulcerative syndrome (EUS) of Asian fishes. Journal of Fish Diseases 16, 169-183.

Rodgers, LJ & Burke, JB (1981) Seasonal variation in the prevalence of 'red spot' disease in estuarine fish with particular reference to the sea mullet, *Mugil cephalus* L. *J Fish Dis,* 4, 297-307.
[http://dx.doi.org/10.1111/j.1365-2761.1981.tb01137.x]

Rukyani, A (1994) Status of epizootic ulcerative disease in Indonesia. In: Roberts, R.J., Campbell, B., MacRae, I.H., (Eds.), *Proceedings of the ODA Regional Seminar on Epizootic Ulcerative Syndrome.* 25-27 January 1994Aquatic Animal Health Research Institute, Bangkok..

Saitanu, K., Wongsawang, S., Sunyascotcharee, B. and Sahaphong, S. (1986) Snakehead fish virus isolation and pathogenicity studies. In: Maclean, J.L., Dizon, L.B. and Hosillos, L.V. (Eds.) The First Asian Fisheries Forum pp. 327-330. Asian Fisheries Society, Manila.

Schalch, S (2021) Deep learning applications for disease diagnostics in aquatic species. *J Aquat Anim Health,* 33, 22-33.

Singh, M & Kaur, G (2023) Evaluating aquafeed additives with AI to prevent EUS in aquaculture systems. *Aquafeed Science and Technology,* 129, 101039.

Stentiford, GD (2020) Emerging technologies to prevent and manage aquatic animal diseases. *Trends Biotechnol,* 38, 576-89.

Sammut, J, White, I & Melville, MD (1996) Acidification of an estuarine tributary in eastern Australia due to drainage of acid sulfate soils. *Mar Freshw Res,* 47, 669-84.

[http://dx.doi.org/10.1071/MF9960669]

Sanjeevaghosh, D (1991) EUS ravages Kerala inland fisheries. *Fishing Chimes,* 11, 47-9.

Shanor, L & Saslow, HB (1944) Aphanomyces as a fish parasite. *Mycologia,* 36, 413-5. [http://dx.doi.org/10.1080/00275514.1944.12017563]

Shariff, M & Law, AT (1980) An incidence of fish mortality in Bekok River, Johore, Malaysia. *Proceedings of the International Symposium on Conservation Input from Life Sciences,* 27-30 October 1980153-62.

Shariff, M & Saidin, TH (1994) Status of epizootic ulcerative syndrome in Malaysia since 1986. In: Roberts, R.J., Campbell, B., MacRae, I.H., (Eds.), *Proceedings of the ODA Regional Seminar on Epizootic Ulcerative Syndrome* 25-27 January 199448.Aquatic Animal Health Research Institute, Bangkok.

Sharifpour, I. (1997) Histology of the inflammatory response of the carp (*Cyprinus carpio* L.) to various stimuli. Ph.D. Thesis, University of Stirling, Scotland.

Shrestha GB. Status of epizootic ulcerative syndrome (EUS) and its effects on aquaculture in Nepal. In: Roberts RJ, Campbell B, MacRae IH, editors. Proceedings of the ODA Regional Seminar on Epizootic Ulcerative Syndrome; 1994 Jan 25-27; Bangkok. Bangkok: Aquatic Animal Health Research Institute; 1994. p. 49-57.

Smith, D. and Onions, A.H.S. (1994) The preservation and maintenance of living fungi. International Mycological Institute, Technical Handbooks No.2. 122. [http://dx.doi.org/10.1079/9780851989020.0000]

Soe, U.M. (1990) Myanmar report. In: Phillips, M. J. and Keddie, H. G., (Eds) Regional Research Programme on Relationships Between Epizootic Ulcerative Syndrome in Fish and the Environment. A Report on the Second Technical Workshop, 13-26 August 1990. pp. 35-38 Network of Aquaculture Centres in Asia-Pacific, Bangkok.

Srivastava, RC (1979) Aphanomycosis-a new threat to fish population. *Mycoses,* 22, 25-9. [http://dx.doi.org/10.1111/j.1439-0507.1979.tb01677.x] [PMID: 763250]

Subasinghe, RP (1993) Effects of controlled infections of *Trichodina* sp. on transmission of epizootic ulcerative syndrome (EUS) to naive snakehead, *Ophicephalus striatus* Bloch. *J Fish Dis,* 16, 161-4. [http://dx.doi.org/10.1111/j.1365-2761.1993.tb00861.x]

Subramaniam, S., Chew-lim, M., Chong, S.Y., Howe, J., Ngoh, G.H. and Chan, Y.C. (1993) Molecular and electron microscopic studies of infectious pancreatic necrosis virus from snakehead. In: IXth International Congress of Virology. Glasgow, UK (Abstract), P. 368.

Suthi, G. (1991) Pathogenicity of motile aeromonads for Puntius schwanfeldi and Oreochromis niloticus with particular reference to the ulcerative disease syndrome (EUS). M.Sc. Thesis, University of Stirling, Scotland. 71.

Tangtrongpiros, J, Charoennetisart, P, Wongsatayanont, B, Tavatsin, A & Chaisiri, N (1985) Haematological change in snakehead fish during 1984 outbreak. *Proceedings of the Technical Conference on Living Aquatic Resources.* 7-8 March 1985Chulalongkorn University, Bangkok..

Thompson, KD, Lilley, JH, Chinabut, S & Adams, A (1997) The antibody response of snakehead,Channa striataBloch, toAphanomyces invaderis. *Fish Shellfish Immunol,* 7, 349-53. [http://dx.doi.org/10.1006/fsim.1997.0084]

Thompson, KD, Lilley, JH, Chen, S-C, Adams, A & Richards, RH The immune response of rainbow trout

(Oncorhynchus mykiss) against Aphanomyces invadans. *Fish Shellfish Immunol* in press.

Tonguthai, K (1985) *A preliminary account of ulcerative fish diseases in the Indo-Pacific region (a comprehensive study based on Thai experiences).* National Inland Fisheries Institute, Bangkok.

Torres, JL, Shariff, M & Law, AT (1990) Identification and virulence screening of Aeromonas spp isolated from healthy and epizootic ulcerative syndrome (EUS)-infected fish. In: Hirano, R., Hanyu, I., (Eds.), *Proceedings of the Second Asian Fisheries Forum* Tokyo, Japan17-22 April 1989663-7.Asian Fisheries Society, Manila.

Ulcerative Fish Disease Committee. (1983) Practical Report of the Ulcerative Fish Disease Committee 1982-1983, Bangkok.

Valairatana, W & Willoughby, LG (1994) The aquatic fungi Aphanomyces and Pythium, as wound pathogens on a soft shell turtle (Trionyx cartilogineus). *AAHRI Newsletter,* 3, 2.

Virgona, JL (1992) Environmental factors influencing the prevalence of a cutaneous ulcerative disease (red spot) in the sea mullet, *Mugil cephalus* L., in the Clarence River, New South Wales, Australia. *J Fish Dis,* 15, 363-78.
[http://dx.doi.org/10.1111/j.1365-2761.1992.tb01236.x]

Vishwanath, TS, Mohan, CV & Shankar, KM (1997) Mycotic granulomatosis and seasonality are the consistent features of epizootic ulcerative syndrome of fresh and brackishwater fishes of Karnataka, India. *Asian Fish Sci,* 10, 155-60.
[http://dx.doi.org/10.33997/j.afs.1997.10.2.007]

Vishwanath, TS, Mohan, CV & Shankar, KM (1998) Epizootic Ulcerative Syndrome (EUS), associated with a fungal pathogen, in Indian fishes: histopathology—'a cause for invasiveness'. *Aquaculture,* 165, 1-9.
[http://dx.doi.org/10.1016/S0044-8486(98)00227-0]

Viswanath, TS, Mohan, CV & Shankar, KM (1997) Clinical and histopathological characterization of different types of lesions associated with epizootic ulcerative syndrome (EUS). *J Aquacult Trop,* 12, 35-42.

Wattanavijarn, W, Torchy, C, Tangtrongpiros, J & de Kinkelin, P (1988) Isolation of a birnavirus belonging to Sp serotype, from southeast Asia fishes. *Bull Eur Assoc Fish Pathol,* 8, 106-8.

Widagdo, D. (1990) Indonesia report. In: Phillips, M.J. and Keddie, H.G. (Eds) Regional Research Programme on Relationships Between Epizootic Ulcerative Syndrome in Fish and the Environment. A Report on the Second Technical Workshop, 13-26 August 1990. Network of Aquaculture Centres in Asia-Pacific, Bangkok.

Willoughby, LG (1995) Aphanomyces invaderis, the fungal pathogen of EUS. C/ N ratios and morphogenesis. *AAHRI Newsletter,* 4, 1-2.

Willoughby, LG & Chinabut, S (1996) Self-staling in Aphanomyces invaderis, the fungal pathogen of freshwater, tropical fish affected by epizootic ulcerative syndrome (EUS). *AAHRI Newsletter,* 5, 2-3.

Willoughby, LG & Lilley, J (1992) The ecology of aquatic fungi in Thailand, and the fish disease relationship. *AAHRI Newsletter,* 1, 5-6.

Willoughby, LG & Roberts, RJ (1994) Improved methodology for isolation of the *Aphanomyces* fungal pathogen of epizootic ulcerative syndrome (EUS) in Asian fish. *J Fish Dis,* 17, 541-3.
[http://dx.doi.org/10.1111/j.1365-2761.1994.tb00250.x]

Willoughby, LG, Pickering, AD & Johnson, HG (1984) Polycell-gel assay of water for spores of

Saprolegniaceae (fungi), especially those of the Saprolegnia pathogen of fish. *Hydrobiologia,* 114, 237-48. [http://dx.doi.org/10.1007/BF00031875]

Willoughby, LG, Roberts, RJ & Chinabut, S (1995) *Aphanomyces invaderis* sp. nov., the fungal pathogen of freshwater tropical fish affected by epizootic ulcerative syndrome. *J Fish Dis,* 18, 273-6. [http://dx.doi.org/10.1111/j.1365-2761.1995.tb00302.x]

Wilson KD, Lo KS. Fish disease in Hong Kong and the potential role of the veterinarian. In: 23rd Annual Conference of the International Association for Aquatic Animal Medicine; 1992 May 18-22; Hong Kong. 1992.

Xuan, T.T. (1990) Vietnam report. In: Phillips, M.J. and Keddie, H.G. (Eds) Regional Research Programme on Relationships Between Epizootic Ulcerative Syndrome in Fish and the Environment. A Report on the Second Technical Workshop, 13-26 August 1990. Network of Aquaculture Centres in Asia-Pacific, Bangkok.

Yu, X (2022) Integrating image-based deep learning with molecular markers for early EUS detection. *Front Vet Sci,* 9, 836542.

Yadav, M, Indira, G & Ansary, A (1992) Cytotoxin elaboration by *Aeromonas hydrophila* isolated from fish with epizootic ulcerative syndrome. *J Fish Dis,* 15, 183-9. [http://dx.doi.org/10.1111/j.1365-2761.1992.tb00652.x]

Zhang, Y (2020) Optimization of water quality management in aquaculture using reinforcement learning algorithms. *Water Res,* 171, 115393.

Zhou, W (2019) Automated surveillance systems for fish health using AI and IoT. *Comput Electron Agric,* 162, 79-89.

SUBJECT INDEX

A

B

C

D

Podeti Koteshwar Rao

E

F

G

H

I

K

L

M

N

O

P

Q

R

S

T

www.ingramcontent.com/pod-product-compliance
Lightning Source LLC
LaVergne TN
LVHW070949210926
848482LV00002B/151
9798898815363